Y0-BDZ-914

Laser Weapons in Space

Westview Replica Editions

The concept of Westview Replica Editions is a response to the continuing crisis in academic and informational publishing. Library budgets for books have been severely curtailed. Ever larger portions of general library budgets are being diverted from the purchase of books and used for data banks, computers, micromedia, and other methods of information retrieval. Interlibrary loan structures further reduce the edition sizes required to satisfy the needs of the scholarly community. Economic pressures on the university presses and the few private scholarly publishing companies have severely limited the capacity of the industry to properly serve the academic and research communities. As a result, many manuscripts dealing with important subjects, often representing the highest level of scholarship, are no longer economically viable publishing projects--or, if accepted for publication, are typically subject to lead times ranging from one to three years.

Westview Replica Editions are our practical solution to the problem. We accept a manuscript in camera-ready form, typed according to our specifications, and move it immediately into the production process. As always, the selection criteria include the importance of the subject, the work's contribution to scholarship, and its insight, originality of thought, and excellence of exposition. The responsibility for editing and proofreading lies with the author or sponsoring institution. We prepare chapter headings and display pages, file for copyright, and obtain Library of Congress Cataloging in Publication Data. A detailed manual contains simple instructions for preparing the final typescript, and our editorial staff is always available to answer questions.

The end result is a book printed on acid-free paper and bound in sturdy library-quality soft covers. We manufacture these books ourselves using equipment that does not require a lengthy make-ready process and that allows us to publish first editions of 300 to 600 copies and to reprint even smaller quantities as needed. Thus, we can produce Replica Editions quickly and can keep even very specialized books in print as long as there is a demand for them.

About the Book and Editor

Laser Weapons in Space: Policy and Doctrine

edited by Keith B. Payne

This is the first comprehensive examination of the issues surrounding the potential development by the United States of a space-based laser weapons program. The authors assess the implications of arms control agreements for a satellite-based laser program, including discussions of recent Soviet space-related arms control initiatives and the forthcoming ABM treaty review. They outline likely Soviet responses to a U.S. space-based laser system, address criticisms of the proposed program, and consider its future in light of developments in U.S. defense strategy and doctrine.

Keith B. Payne is vice-president and director of national security studies at the National Institute for Public Policy and author of *Nuclear Deterrence in U.S.-Soviet Relations* (Westview, 1982).

Laser Weapons in Space

Policy and Doctrine

edited by Keith B. Payne

Westview Press / Boulder, Colorado

A Westview Replica Edition

Published in 1983 in the United States of America by
Westview Press, Inc.
5500 Central Avenue
Boulder, CO 80301
Frederick A. Praeger, President and Publisher

Library of Congress Cataloging in Publication Data
Main entry under title:
Laser weapons in space.
(A Westview replica edition)
Includes bibliographical references.
1. Space warfare--Addresses, essays, lectures.
2. Lasers--Military applications--Addresses, essays, lectures. 3. Arms control--Addresses, essays, lectures. I. Payne, Keith B.
UG1530.L37 1983 358'.8 82-20045
ISBN 0-86531-937-5 (pbk.)

10 9 8 7 6 5 4 3 2

Contents

Tables and Figures

Tables

Figures

Acknowledgments

I would like to thank Dan Strode of the National Institute for Public Policy for his contributions to this book. In addition, I would like to thank Candace Smoot, Beth Miller and Sue Munn for their invaluable and patient assistance in the preparation of the manuscript.

Keith B. Payne

1
Introduction and Overview of Policy Issues

Keith B. Payne

This book addresses the policy issues associated with space-based lasers as a means of ballistic missile defense (BMD). It is not a technical feasibility study, and no attempt has been made to calculate either the absolute or the opportunity costs of space-based lasers. However, the following chapter by Dr. Patrick Friel, does provide a technical overview of space-based BMD. The focus of the book is on a different order of question: how well or poorly would such a system (assuming it could be built and built effectively) serve the national interest? More specifically, the implications of space-based lasers for arms control, U.S.-Soviet relations, deterrence, and nuclear strategy, are examined in detail.

In this chapter it is argued that space-based lasers would enhance, not undermine, crisis stability. Space-based lasers could be used to improve the survivability of U.S. strategic forces, promoting deterrence. It is the assured survivability of a retaliatory force, not the vulnerability of our own society, which is the key to deterrence stability. It is argued that by promoting deterrence, space-based lasers would accomplish one of the primary objectives of arms control -- to reduce the probability of war. While deployment of a space-based BMD system would involve either renegotiation or withdrawal from the ABM Treaty, the strategic rationale for continued U.S. support of that agreement is not unassailable. The failure to achieve an adequate arms control agreement on offensive arms which would ensure the survival of land-based ICBMs, has removed much of the rationale for continued acceptance of the Treaty.

In the third chapter, "Implications of Arms Control Agreements for Space-Based Laser BMD," Alan Jones analyzes those provisions of the 1972 ABM Treaty and other international agreements which impinge on the

A much expanded version of this chapter appeared in the fall 1982 issue of International Security Review.

development, testing, and deployment of space-based lasers. Dr. Jones argues that the ABM Treaty is the only extant agreement that would affect a decision to pursue this type of technology. However, the Treaty permits "substantial partial testing of components for space-based BMD lasers while prohibiting full-system testing and deployment of such systems."

In chapter four, "Space-Based Lasers for Ballistic Missile Defense: Soviet Policy Options," Rebecca Strode examines possible Soviet military responses to U.S. space-based laser BMD, the role of missile defense in Soviet strategy, and the manner in which the U.S.S.R. has responded to technological challenges in the past. Ms. Strode concludes that there is little chance that the Soviet Union would respond to deployment of a U.S. space-based laser with a direct attack. Instead, the Soviets would use political means to drive a wedge between Western Europe and the United States, arguing that American actions were responsible for "the death of arms control." Ms. Strode also analyzes a number of technical responses, including comparable BMD deployments and possible increases in offensive systems. While the Soviets would probably wish to combine a number of potential responses, economic constraints may force greater reliance on one option in the long run. Ms. Strode speculates that this option would probably be BMD.

The essential backdrop for the policy issues sparked by space-based lasers is the on-going debate over strategy. Barry Schneider traces the evolution of U.S. nuclear strategy since the Second World War in chapter five, "Space-Based Lasers and the Evolution of Strategic Thought." The combination of nuclear weapons with ballistic missiles caused a decisive shift in favor of the offense over the defense. This led to strategic theories such as Mutual Assured Destruction (MAD), which based deterrence solely on the ability to retaliate against the opponent's urban/industrial base. Starting with the Schlesinger Doctrine in 1974, there has been a continuous shift towards strategies which rest deterrence on the ability to neutralize any military advantages which the Soviet Union might gain by attacking. In this context, there has been renewed interest in strategic defense. Space-based laser BMD holds out some hope (although there are admitted technical obstacles) that the balance between the offense and the defense might be restored.

In the concluding chapter, Colin S. Gray surveys the Reagan Administration's defense policy with particular emphasis on its attitudes towards strategic defense. The Administration has shown greater interest in strategic defense than its predecessor, demonstrated by increased funding levels. However, many in the Administration are skeptical about the technical possibilities and fearful of a negative public reaction. This is unfortunate since the greatest gap in current American strategy is its failure to provide for the defense of our population. If space-based laser BMD is feasible, Dr. Gray argues, then it could make a valuable contribution to our deterrence posture by serving as the first layer of defense. Dr. Gray cautions, however, that lasers will not remove the need for other strategic systems, but will be one part of a layered force structure combining both offensive and defensive elements.

DETERRENCE STABILITY

It must be recognized that deterrence stability is the primary U.S. objective in its relations with the Soviet Union. The framework within which space-based lasers must be regarded is that of deterrence. Consequently, an issue of critical import is, "how does a space-based laser program contribute to the deterrence objective; and would a space-based laser program degrade deterrence?"

The primary non-technical criticism directed toward space-based laser development is that such a system, if ever deployed, would promote crisis instability. Crisis instability involves the "reciprocal fears of preemption" that would intensify during an intense political/military crisis. According to this theory, reciprocal fears of preemption would cause mutual incentives to preempt, to launch the first-strike rather than suffer the first-strike, and obtain military benefits therein. A space-based laser BMD system would promote "crisis-instability," according to its critics, because it could lead American decision-makers to think that a nuclear war could be waged successfully: that is, the U.S. could launch a first-strike and thereby degrade Soviet retaliatory capabilities, and then degrade the effectiveness of Soviet retaliatory strikes via employment of space-based laser BMD. Thus, the U.S. could deny the Soviet Union its deterrent, and the Soviet leadership, fearing precisely such an occurrence during a crisis, would seek to gain the advantage of the first-strike. In terms of Soviet strategic forces, the Soviet Union would be placed under pressures to "use'em or lose'em", as the jargon has it.

The U.S. has long taken the prospect of crisis instability very seriously, and to a great extent, it has deployed forces that would, in principal, promote crisis stability.[1] As former Secretary of Defense Harold Brown observed:

> In the interests of stability, we avoid the capability of eliminating the other side's deterrent, insofar as we might be able to do so. In short, we must be quite willing--as we have been for some time--to accept the principle of mutual deterrence and design our defense posture in light of that principle. (Emphasis added)[2]

There is nothing in previous Soviet behavior or strategic doctrine to suggest that any specific weapons, such as space-based lasers, would be targeted apart from a more general war; and, if they were not willing to risk a general war, attacks on U.S. space-based lasers would make little sense. Only if the Soviets feared that the United States was on the verge of completing a decisive war-winning weapon in space that would totally change the strategic balance to their disadvantage, would preemptive attacks on U.S. SBL-BMDs make any sense. Even then, unless the Soviet civil, air, and missile defenses were capable of neutralizing U.S. retaliatory nuclear attacks, such an idea would be inviting disaster.

Nevertheless, leaders of powerful states do not easily stand by and watch their adversaries gain a strategic military lead over them. The 1962 Cuban Missile Crisis was one indication of the severe reaction of a superpower, the United States, not permitting the other superpower to gain a 5 to 10 year advantage (allowing them to "catch up" strategically) by deploying new weapons (MRBMs and IRBMs) within range of targets then only partially covered by Soviet ICBMs. The Israeli air attack on Iraq's nuclear reactor facility in Osirak on June 7, 1981, is a recent example of a state willing to attack preemptively a facility that could produce a decisive shift in the balance of power. Might not the building of space-based laser battle stations invite a high altitude "Osirak"? The Soviets might decide to risk it figuring that the loss of U.S. life would be minimal and that the events would take place in space, far from U.S. territory. As Edgar Ulsamer has observed:

> Those who shrug off the vast technological and engineering problems associated with a

> .....space-based laser ballistic missile defense system still confront a monumental obstacle: the excessive vulnerability of such a space battle station combined with the excessive incentive for the other side to attack it preemptively.[3]

While such preemptive tactics have no precedent in Soviet behavior, and while this would invite a costly U.S. retaliation, if the U.S. space-based laser system was threatening enough, Soviet leaders still might strike out into uncharted skies and launch a preemptive Pearl Harbor in space. Since highly efficient SBL-BMDs might one day shift the strategic balance, they would serve as tempting targets before they could be completed.

It does not, however, seem likely that a space-based laser BMD system would promote crisis instability. The U.S. has limited capabilities to attack hardened Soviet ICBMs. A first-strike by U.S. missiles against Soviet silos would destroy no more than 30 percent of the modern Soviet SS-17, SS-18, and SS-19 launchers.[4] As mentioned above, the U.S. consciously has avoided force modernization programs that would threaten the Soviet deterrent. No space-based laser program under consideration could come near to coping with the number of Soviet ICBMs that would survive a U.S. first-strike. Space-based laser BMD would not deny the Soviet Union its deterrent.

Modernization of U.S. strategic forces is likely to add countersilo competence to the U.S. strategic force posture. Consequently, it could be argued that a space-based laser BMD program, complemented by 2000 countersilo warheads, would constitute a first-strike threat in the Soviet perspective, and thus promote crisis instability.

However, modernization of the U.S. ICBM force will entail enhanced survivability. It is now fairly obvious that if the MX is deployed, it will have to be done in what at least appears to be a highly survivable basing mode. Additionally, modernization of the strategic bomber leg of the triad will emphasize survivability with the deployment of the B-1B and possibly a "Stealth" (Advanced Technology Bomber) follow-on. It has long been understood that strategic force survivability is the key to minimizing crisis instability, regardless of the degree of vulnerability of urban and industrial assets. If U.S. strategic forces possess a high level of survivability, and thus the U.S. retaliatory threat cannot be degraded by a Soviet first-strike, crisis instability should not occur.

Crisis instability would not emerge because the Soviet Union could not anticipate minimizing losses or gaining an advantageous position through preemption. Indeed, a very high degree of survivability should ensure a net loss for the Soviet Union if it launched a first-strike in terms of the "correlation of forces". Such a situation would minimize crisis instability, even if the U.S. were capable of significant damage-limitation in nuclear war.

The Soviet perspective on this matter is extremely important. The Soviet Union has exhibited very little concern in promoting crisis instability in its buildup of a formidable countersilo capability. Former Secretary of Defense Harold Brown observed in 1980 that the U.S. ICBM force was then, or soon would be, vulnerable to a Soviet first-strike.[5] Soviet leaders obviously are less concerned than their American counterparts in "threatening the opponent's deterrent."

The Soviets have, however, enforced crisis stability by ensuring the survivability of their own strategic forces; something the U.S., with all its officially expressed concern for crisis instability, has failed to do. The Soviet effort to enforce crisis stability is instructive when considering the case made against space-based laser BMD because it illustrates well the point made above: providing U.S. strategic forces with a high degree of survivability is the key to stability, not maintaining the absolute vulnerability of urban/industrial America.

Another Soviet perspective that is important in this regard is their view of the "correlation of forces" and the objectives of the initial phase of nuclear war. The Soviet approach to nuclear war specifies that during the initial phase of operations, a "decisive" change in the correlation of forces must be achieved.[6] Soviet writing on nuclear war is clear in its focus upon the need for preemption, surprise, and primarily counterforce operations against American military forces, particularly nuclear forces.[7] The Soviets emphasize that to engage in nuclear threat without such a capability would be adventurism of the worst order.[8]

What are the implications of this Soviet perspective for space-based laser BMD, and the allegation that it would promote crisis instability? First, degrading the Soviet capability to retaliate against American cities is unlikely to enhance Soviet incentives to preempt. Thus, if space-based laser BMD, in conjunction with enhanced U.S. counterforce capabilities, poses a threat to Soviet urban/industrial retaliatory capabilities, Soviet incentives to preempt in a crisis need not heighten. Second, the surest means of minimizing Soviet incentives to preempt would be to provide for the survivability of U.S. strategic forces; doing so would ensure that the Soviet Union could not "change the correlation of forces" via a first-strike. Consequently, to minimize crisis instability, the U.S. should concentrate upon force survivability, not societal vulnerability. A reduction in vulnerability of urban/industrial America should not affect Soviet considerations of preemption as long as U.S. strategic forces are survivable. Thus, a highly survivable U.S. strategic force posture may be considered necessary for crisis stability not societal vulnerability, (assuming, of course, the penetrativity of U.S. forces in addition).

The necessity for highly survivable strategic forces to address the potential problem of crisis instability is critical to the development of weaponized, space-based lasers. This is because the system(s) available during the foreseeable future would be more appropriate for providing force survivability than reducing urban/industrial vulnerability. Consequently the case can, and indeed should, be made that space-based laser development would not pose a potential threat to crisis stability; rather, if it can perform the missions envisaged, such as preferential ICBM launcher defense, _it would contribute to crisis stability_.

DISARMAMENT AND ARMS CONTROL

As is the case with other strategic weapons programs, or potential programs, intended to enhance the survivability of U.S. deterrent forces, (e.g. _Trident_ ballistic missile carrying submarines, B-1B _bombers_, conventional BMD for ICBM defense, hardened ICBM launchers, etc.), augmenting such U.S. capabilities _should reduce the probability of war._ That is, promoting U.S. strategic force survivability should lend credence to the American retaliatory threat and minimize crisis instability. Consequently, it should also reduce the probability of war.

It should be remembered that the two priority objectives of arms control are to:

* reduce the probability of war;

* reduce the level of destruction should war occur.

Thus, development and deployment of space-based lasers for the purpose of protecting U.S. strategic forces should promote the objectives of arms control if the system(s) prove technically and operationally credible. Too often the simple-minded notion is presented that limiting weapons systems indiscriminately promotes arms control and reduces the probability of war. That position, though popular and emotionally satisfying, is very likely to be wrong. Those systems designed to enhance U.S. force survivability should _reduce_ the probability of war by enhancing deterrence stability; to limit or ban those systems would, in effect, contribute to the degradation of stability and thereby _increase_ the probability of war. Simple-minded efforts to control weapons can be dangerous, particularly if they are not assessed very carefully; that indeed is the problem with efforts to "freeze" all strategic weapons systems, or ban the military exploitation of space. For example, to hold the current strategic balance frozen in place now would leave two legs of the U.S. strategic force triad (land based ICBMs, and strategic bombers) highly vulnerable, an extremely dangerous and destabilizing prospect.

It must be recognized that promoting stability and reducing the probability of war is the ultimate objective. Efforts to control strategic forces are intended to serve those objectives; arms control is not an objective in and of itself. Ironic as it appears, those sincere efforts to "freeze" weapons, and hold space as a "zone of peace" could easily be counterproductive in terms of promoting the objectives of arms control. To minimize the probability of war the U.S. needs to act now to reduce the high degree of strategic force vulnerability that has resulted from the formidable buildup of Soviet counterforce potential. To the degree that the development and deployment of space-based laser technology for BMD would serve that goal, it would also promote the primary objective of arms control, i.e. reducing the probability of war.

In addition, it is quite conceivable that the development of space-based laser technology will stimulate Soviet interest in arms control as the appropriate response. The ABM Treaty analogy is illustrative. The Soviet interest in serious strategic arms control became effective within days following congressional approval of _Sentinel_. Soviet interest in limiting the American BMD system was paramount throughout the SALT I negotiations. Indeed, once mutual agreement on BMD limitation was clear, the Soviets were anxious to conclude an agreement that avoided offensive force limitations entirely. The Soviet Union obviously sought to maximize its counterforce/countersilo potential by limiting the U.S. BMD program, and by leaving unfettered the potential counterforce capabilities residing in the then ongoing development of fourth-generation SS-19 and SS-18 ICBM systems. The point is that the Soviets agreed to the ABM Treaty largely to maintain unimpeded targeting access to U.S. ICBMs and NCA facilities and to counter the U.S. BMD technological advantage. The only reason offensive force limitations were effected was because the U.S. required the linkage between offensive and defensive limitations. Consequently, historical evidence points to the fact that the incentive that has brought the Soviet Union to the negotiating table in the past has been the result of U.S. weapons programs. Further, the incentive that has proven particularly effective has been the development of a weapons system that threatened the Soviet capability to target U.S. ICBMs. This is because Soviet strategic doctrine emphasizes as <u>the</u> essential aspect of the Soviet "theory of victory" the destruction of American nuclear forces during the initial phase of nuclear war.

In short, an American capability to impede Soviet targeting of U.S. retaliatory capabilities threatens Soviet strategy. In the past, the Soviets have turned to arms control negotiations as the response to such a development, i.e. the SALT I negotiations and the resultant ABM Treaty. In short, given past experience the development of space-based laser technology suitable for enhancing U.S. strategic force survivability would be likely to enhance Soviet motivations to engage in serious arms control negotiations. It is quite clear that the Soviet Union will not engage seriously in strategic arms control negotiations without such incentives.

THE ABM TREATY

As discussed in chapter three of this book, the U.S. must revise, or withdraw from the ABM Treaty in order to deploy a space-based laser system for BMD. However, many in the U.S. arms control community consider the ABM Treaty to be the most effective existing arms control agreement, and would oppose strongly any move to revise or withdraw from the Treaty. Additionally, it would appear to much of the interested public that the U.S. is opposed to arms control if it attempted to alter the status of the ABM Treaty (with all of the propaganda value that would offer the Soviet Union). Consequently, there would be significant opposition to any effort to revise or withdraw from the ABM Treaty.

Advocates of the ABM Treaty believe that scrapping it and deploying ABMs would be a mistake. They argue that peace can more surely be preserved by leaving both sides defenseless rather than putting the retaliatory capability of our strategic forces into doubt. As Winston Churchill once observed:

> It may well be that we shall by a process of supreme irony have reached a stage in this story where safety will be the sturdy child of terror, and survival the twin brother of annihilation."[9]

Defenders of the ABM Treaty emphasize the fact that under current conditions U.S. strategic systems can reach their intended targets should our forces be attacked. The Soviets know this. Thus, they are deterred. This is the logic of the balance of terror. ABM Treaty supporters also argue that:

-- the agreement preserves the much weaker strategic deterrents of the United Kingdom, France, and China.

-- the Soviet Union might acquire a viable ballistic missile defense before the United States could achieve one without the restraints of the Treaty.

-- abrogation of the ABM Treaty would trigger a new arms race between offensive and defensive strategic weapons that would jeopardize stability, cost hundreds of billions, and undercut strategic arms reduction efforts.

-- the U.S. triad (SAC Bombers, ICBMs, SLBMs) has an assured destruction capability to deter Soviet attack and, despite vulnerability of ICBMs, the U.S. has adequate retaliatory capability even without ABM protection of U.S. ICBMs.

-- the tens of billions of U.S. dollars necessary for R&D, production, and deployment of ABMs might be better used to improve the strategic forces and readiness of U.S. and NATO conventional forces.

-- offensive missile technology enjoys commanding technological advantages over present or forseeable BMD technology, and far cheaper countermeasures could negate extremely expensive new BMD efforts.

Probably the most prevelant argument against ballistic missile defenses is that there is serious doubt concerning whether they can be made to work with sufficient efficiency.

This argument was made fifteen years ago by then Secretary of Defense Robert McNamara:

> ...it is important to understand that none of the (ABM) systems at the present or foreseeable state-of-the-art would provide an impenetrable shield over the United States. Were such a shield possible, we would certainly want it -- and we would certainly build it. If we could build and deploy a genuinely impenetrable shield over the United States, we would be willing to spend not $40 billion (in 1967 dollars) but any reasonable multiple of that amount that was necessary. The money in itself is not the problem; the penetrability of the proposed shield is the problem.[10]

However, the search for the perfect can become the enemy of good enough. McNarara's ABM standard of adequacy probably was much too high. Point defenses, (e.g., the defense of ICBMs) can contribute to stability and deterrence, even if perfect area defenses are beyond the scope of present BMD technologies.

The U.S. has been forced by the Soviet strategic nuclear buildup of the 1970s to reconsider its commitment to the Treaty. Since the signing of the SALT I agreement in 1972, the Soviet Union has increased its counterforce/countersilo capability to the point where American ICBMs are now almost entirely vulnerable. Revising the ABM Treaty may be the only means of providing a high degree of survivability for the land-based ICBM force. Such an action would not reflect American disdain for arms control. Rather, it would reflect a commitment to maintain deterrence stability because, as discussed above, deterrence stability requires strategic force survivability.

At the time of the signing of the ABM Treaty, the U.S. recognized a very clear linkage between offensive strategic force limitations and the ABM Treaty. The U.S. consciously established a linkage between offensive and defensive limitations. The notion guiding that linkage was to the effect that offensive force limitations would constrain the Soviet counterforce/countersilo capability such that the U.S. ICBM force would not require BMD coverage to provide survivability. Thus, the U.S. hoped that the Interim Offensive Force Agreement (SALT I, 1972) would allow the U.S. to give up the BMD option and maintain a survivable ICBM force by constraining the Soviet counterforce threat. Additionally it was assumed that future, more significant, offensive force limitations would follow SALT I and further constrain Soviet capabilities, thus ensuring force survivability. Indeed, in Unilateral Statement A the leader of the U.S. SALT delegation, Ambassador Smith, stated that if a more comprehensive follow-on agreement limiting offensive forces was not attained within five years, thus providing for U.S. force survivability, the U.S. would consider it to be an extraordinary event within the framework of the conditions required for the U.S. to reconsider its commitment to the Treaty. The withdrawal clause in Article XV of the ABM Treaty reads:

> Each Party shall, in exercising its national sovereignty, have the right to withdraw from this Treaty if it decides that extraordinary events related to the subject matter of this Treaty have jeopardized its supreme interests. It shall give notice of its decision to the other Party six months prior to withdrawal from the Treaty. Such notice shall include a statement of extraordinary events the notifying Party regards as having jeopardized its supreme interests.

In Unilateral Statement A, Ambassador Smith formalized this linkage in offensive and defensive force limitations as part of American arms control policy:

> The U.S. Delegation has stressed the importance the U.S. Government attaches to achieving agreement on more complete limitations on strategic offensive arms, following agreement on an ABM Treaty and on an Interim Agreement on certain measures with respect to the limitation of strategic offensive arms. The U.S. Delegation believes that an objective of the follow-on negotiations should be to constrain and reduce on a long-term basis threats to the survivability of our respective strategic retaliatory forces. The U.S.S.R. Delegation has also indicated that the objectives of SALT would remain unfulfilled without the achievement of an agreement providing for more complete limitations on strategic offensive arms.

> Both sides recognize that the initial agreements would be steps toward the achievement of more complete limitations on strategic arms. If an agreement providing for more complete strategic offensive arms limitations were not achieved within five years, U.S. supreme interests could be jeopardized. Should that occur, it would constitute a basis for withdrawal from the ABM Treaty. The U.S. does not wish to see such a situation occur, nor do we believe that the U.S.S.R. does. It is because we wish to prevent such a situation that we emphasize the importance the U.S. Government attaches to achievement of more complete limitations on strategic offensive arms. The U.S. Executive will inform the Congress, in connection with Congressional consideration of the ABM Treaty and the Interim Agreement, of this statement of the U.S. position.

However, the anticipated more comprehensive offensive force limitation was not achieved within five years, i.e. SALT II was not signed until 1979 (and, of course, never ratified by the U.S.). More importantly, neither SALT I nor SALT II denied the Soviet Union the counterforce/countersilo potential necessary to target the U.S. ICBM force. Even the proponents of SALT recognize that fact. Thus the very reasonable conditions established by the U.S. in terms of offensive force limitations and continued support for the ABM Treaty have never been met. Consequently the U.S., by its very arms control policy statements, should seriously review its support of the ABM Treaty; and the dangerous vulnerability of the U.S. ICBM force reinforces the need for such a review. Review of the ABM Treaty would not be an American anti-arms control initiative, it would be forced upon the U.S. by the extensive Soviet strategic buildup. The strategic rationale for the ABM Treaty has been destroyed by the decade-long buildup of Soviet counterforce capabilities, and the lack of success in attaining offensive strategic force arms control.

SUMMARY

In summary, the policy-oriented opposition to any potential space-based laser BMD system(s) will be significant. The policy critiques brought to bear against development of space-based lasers will include charges that such a system would promote crisis instability, hinder the cause of arms control, lead to the militarization of space, and force the termination of the most effective existing arms control agreement, the ABM Treaty. Indeed, these arguments are already presented forthcoming by critics of space-based laser development.

However, each of these criticisms is of questionable validity. It must be remembered that the U.S. objective is indeed deterrence. The pursuit of deterrence is the framework within which the development of space-based laser technology will be assessed. However, that fact should not work against the development of laser technology. Efforts to enhance U.S. strategic force survivability, as is envisioned for any space-based laser system, should, if effective, function to reduce crisis instability and enhance deterrence stability, and consequently reduce the probability of war.

Concerning the arms control critiques, the primary objective of any arms control endeavor is to reduce the probability of war. A space-based laser BMD system should, if effective, actually promote the primary goal of arms control by enhancing the survivability of U.S. strategic forces. "Freezes," limitations, and maintaining "zones of peace," are not goals in and of themselves. They should serve the goals of arms control. If they inhibit that objective by denying the U.S. the capability to ensure force survivability, they must be avoided. In addition, it is quite conceivable that U.S. development of space-based laser BMD technology would motivate the Soviet Union to engage seriously in arms control negotiations. There is little likelihood that it would do so without such incentives.

Finally, the strategic rationale for continued U.S. support for the ABM Treaty has been shattered by over a decade-long buildup of Soviet countersilo capabilities (particularly in the deployment of SS-19s and SS-18s). Indeed, U.S. arms control policy itself now would require a serious reexamination of U.S. support for the ABM Treaty. The very reasonable and strategically sound linkage between offensive and defensive arms control limitations established by the U.S. during SALT I should not be forgotten. In short, the lack of success in offensive force limitations and the current vulnerability of U.S. ICBMs should lead the U.S., under Article XV and Unilateral Statement A of the Treaty, to review its commitment to the Treaty.

Such an action would not reflect an anti-arms control penchant on the part of the U.S., but an effort to ensure the continuation of deterrence stability, and a recognition of the necessity of maintaining a useful integrity between offensive and defensive force limitations.

NOTES

1. For example, the Brooke Amendment ruled out Stellar Inertial Guidance (SIG) on Polaris submarines.

2. Harold Brown, Department of Defense Annual Report, FY1980 (Washington, D.C.: USGPO, Jan. 25, 1979), p. 61.

3. Edgar Ulsamer, "The Long Leap Toward Space Laser Weapons," Air Force Magazine, Aug. 1981, Vol. 64, No. 8, p. 63.

4. See Current News, Part II (April 13, 1982), p. 12-F.

5. Harold Brown, Remarks Prepared for Delivery by the Honorable Harold Brown, Naval War College, Newport, Rhode Island, Aug. 20, 1980, in Department of Defense, News Release, No. 344-80 (Aug. 20, 1980), p. 1.

6. See for example, V.D. Sokolovskiy, Soviet Military Strategy, 3rd editon, translated and edited by Harriet Fast Scott (New York: Crane, Russak & Company, 1975), pp. 289-296; A.A. Grechko, On Guard for Peace and the Building of Communism, in JPRS, No. 54602 (Dec. 2, 1971), pp. 32, 36; and V. Reznochenko, "Features and Methods of the Offense," Soviet Military Review, No. 9 (Sept. 1973), p. 10.

7. See Col. S. Tyushkevick, "The Metholdogy for the Correlation of Forces," Voyennaya Mysl, No. 6 (June 1969), FRD 0008/70 (Jan. 30, 1970), pp. 34, 37; and Maj. General I. Anureyev, "Determining the Correlation of Forces in Nuclear Weapons," Voyennay Mysl, No. 6 (June 1967), FPD 0112/68 (July 11, 1968), pp. 39-40, 42.

8. See for example, Leon Goure, War Survival in Soviet Strategy (Wash., DC: Center for Advanced International Studies, University of Miami, 1976), p. 5.

9. Winston Churchill, speech to the House of Commons, March 1, 1955. See Parliamentary Debates, 5th Series Vol. 537, Col. 1893-1905. Also quoted in Theodore Ropp, War in the Modern World (N.Y.: Collier Books, 1962), p. 404.

10. Robert S. McNamara, Speech to UPI editors in San Francisco, California, Sept. 18, 1967. Reprinted in the Department of State Bulletin, Oct. 9, 1967.

2
Space-Based Ballistic Missile Defense: An Overview of the Technical Issues

Patrick J. Friel

The possibility of using outer space as a basis for the development and deployment of a strategic defense to protect the United States against a ballistic missile attack by the Soviet Union has received much public attention during the past few years. The use of "directed energy weapons" (DEW), i.e., a neutral beam of hydrogen atoms (H°) or an intense beam of infrared, visible, or ultraviolet light from high powered lasers has received particular attention, primarily the latter. Space-based lasers offer the promise of destroying the offensive boosters during launch by depositing a lethal amount of radiant energy in the missile's skin from space, essentially instantaneously, i.e., at the speed-of-light (3×10^8 m/sec). The neutral hydrogen beam operates entirely outside the atmosphere and attempts to neutralize the attack by destroying each reentry vehicle (RV) carrying nuclear weapons. It is generally agreed that if an intense H° beam (several hundreds of million electron volts (MEV)) illuminates a reentry vehicle, the energy deposited will most certainly destroy the reentry vehicles. Neither DEW systems can acquire and track the threatening missiles or RV's autonomously. That is, neither beam weapon is agile and cannot provide the number of beams required to detect and track initially a large Soviet ballistic missile launch. (For example, the phased-array radar in conventional, ground-based ballistic missile defense has this capability). Consequently, both DEW systems require an auxiliary sensor, e.g., an active space-based radar or passive infrared sensor to detect, acquire and track a Soviet ballistic missile launch (an "active" sensor bounces radiation off the target to detect its presence and determine its range while a "passive" sensor detects the radiation emitted by the target and determines its range by tracking its angular motion).

Although the space-based DEWs have received the greatest public attention, there are two other space-based ballistic missile defense concepts which have been considered but have not been widely publicized. In both cases the kill mechanism involved a "conventional" or chemically propellant interceptor armed preferably with a non-nuclear warhead but, with the option to use a small nuclear warhead. In one concept both the defensive missiles and auxiliary sensors are space-based. While in the second, the sensors are space-based and the interceptors are deployed in the continental United States (CONUS). In the case where all defense components are space-based, the sensors would acquire and track the booster, infrared homing interceptors would be launched from the space platform, and the offensive missile destroyed in the launch phase. In the second case, where only the sensors are space-based, the sensors would determine the exact trajectory of each threatening RV. The CONUS-based interceptors would be launched to a point in space given the RV trajectories provided by the space-based sensors. The terminal engagement of each RV would involve an infrared homing interceptor armed with a non-nuclear warhead. The RV and its nuclear warhead would be destroyed by the deposition of the enormous impact energies associated with closing velocities between the interceptor and the RV in the exosphere (10 kilometers per second or more).

This chapter presents an overview of the key technical issues associated with all three space-based ballistic missile defense concepts. Particular emphasis will be placed on space-based laser weapons, however, with the objective of providing the reader with an appreciation of the technical characteristics of the components involved. Potential countermeasures which the Soviet Union could employ to neutralize the effectiveness of a space-based laser weapon also will be described.

SPACE-BASED BMD USING "CONVENTIONAL" COMPONENTS

The most credible space-based BMD would be one which involves the use of components which are within the present or a short term projection of the state-of-the-art in sensors and chemically propelled interceptor missiles. The "High Frontier" study[1] advocated the deployment of both interceptor and sensor system in space (a similar concept called "BAMBI" was studied and rejected in the mid-1960s). The system involved the deployment of 492 space "trucks" in 300 nautical mile orbits each carrying 40 to 45 conventional "carrier vehicles" armed with non-nuclear warheads. An unspecified infrared or radar sensor system would acquire and track a Soviet missile launch and provide the trajectory data required to launch the interceptor missiles. Each carrier vehicle could acquire a velocity of 3000 feet per second with respect to the space "truck." The estimated costs for full deployment were relatively modest, $11 billion, while the system could counter a 1000 missile Soviet attack.

Critics have pointed out, however, that the estimated size and costs of this system are very optimistic. First an interceptor with a 3000 foot per second velocity capability with respect to the bus can only intercept attacking missiles which are in the same orbital plane. In fact, the interceptor can engage Soviet missiles launched only at orbit inclinations of 7° or less. The weight of the interceptor required to engage Soviet missiles at an orbit inclination of 45° would be close to a *Minuteman* ICBM or more. Also, the number of satellites required may be substantially understated. The number of satellites needed is equal to the number of satellites required for single point coverage of the entire earth times the number required to be in the battle at one time. The number of satellites required for single coverage of the entire earth has been determined for a 300 nautical mile orbit to be 60[2] assuming a 1000 missile attack, and at the number of interceptors in each space "truck," stated by proponents, the number of "trucks" required at all times in the battlespace would be about 22. The total number of satellites required, therefore, would be 60 x 22 or 1320. The total costs in reality, therefore, may be 10 to 15 times the value the "High Frontier" proponents suggest. There are also substantial technical issues associated with the space-based homing interceptor. For example, the infrared homing sensor must be able to distinguish between the cool missile skin and the enormous radiation levels emitted by the rocket's plume, which is not a trivial technical problem. Thus, many critics consider the "High Frontier" global ballistic missile defense concept to be technically and economically superficial. However, these same critics in many cases in no way disagree with the broad strategic objectives of the study, i.e., to move the U.S. and the Western Alliance away from an offensive strategy of deterring nuclear war through the possession of a "mutual assured destruction" (MAD) capability to one of strategic defense or "assured survival."

The second space-based BMD concept involves the use of space-based sensors to determine the precise impact point of each RV in a Soviet ballistic missile attack and CONUS-based non-nuclear (or small nuclear) interceptors. The sensors would be infrared satellites in synchronous orbit (about eight total) which would detect and track the Soviet missile early in the trajectory and active, millimeter wave radar satellites in low earth orbits which would precisely track each RV when deployed from the MIRV bus. The CONUS interceptors would be launched to a point in space through which each RV would pass. Each infrared homing interceptor would destroy the RV with a non-nuclear (or small nuclear) warhead. There is little doubt that the infrared and millimeter wave sensors required are within the projected state-of-the-art (the author's bias is toward this space-based BMD system concept). However, critics will point out the deficiencies, i.e., the vulnerability of the space-based sensors to direct attack or to a variety of sensor countermeasures. It is true that the U.S. is investigating a wide spectrum of satellite survival aids, e.g., decoys, "silent" spares, reduced observables, maneuvering, rapid replacement, and even active defense. However, the application of these techniques to improvement in the survivability of U.S. space-based strategic assets, particularly when associated with a BMD system, has not been evaluated to any significant extent.

The various deficiencies associated with the use of "conventional" BMD components in space to neutralize a Soviet ballistic missile attack have led many strategic analysts and technologists to the conclusion that the only feasible space-based BMD for the U.S. in the long run will be a "speed-of-light" weapon, DEW's, i.e., neutral particle beams, and particularly high energy lasers.

SPACE-BASED NEUTRAL PARTICLE BEAM WEAPONS

Since the mid-1960s, the U.S. has maintained a program designed to study the application of particle beam weapons to ballistic missile defense. These beam weapons include charged particle beams (protons and electrons, primarily the latter) and neutral particle beams (hydrogen atoms). Charged particle beams cannot be used in the exosphere since the beam would be deflected by the earth's magnetic field. The possibility of using electron beams in endoatmospheric terminal defense was examined extensively by the U.S. in the 1960s. Those studies indicated that while it may be possible to propagate a lethal beam in the atmosphere, the beam would not be stable. In addition, the applications studies showed that the power required is enormous, the pointing and tracking requirements were severe, and the system would be easily saturated in a large attack. Therefore, it appears that the application of electron beam weapons to endoatmospheric defense is impractical.

In recent years the U.S. has also evaluated the use of neutral particle beams of hydrogen atoms* in space-based ballistic missile defense. Charged particle beam research in the U.S. and the U.S.S.R. has shown that it is feasible to produce an intense collimated beam of H^- particles using available particle accelerator systems. Techniques have been developed to remove the electron without causing the beam to diverge, so that it may be possible to propagate a narrow, intense beam of hydrogen atoms to long distances in the exosphere. The interaction of a high energy H° beam with a reentry vehicle would certainly cause its destruction. However, the application of this neutral beam technology to exospheric ballistic missile defense presents some formidable technological and engineering problems. If it is assumed that a practical range at which a narrow beam could be maintained is about 1000 km, the number of satellites required to ensure coverage of a given missile silo field would be 20 (the great circle length divided by 1000 km). Complete coverage of the U.S.S.R. would require a large number of orbits filled with satellites. Also, if it is assumed that the irradiation times required to produce a lethal dose of protons in the RV is significant, e.g., on the order of 10 seconds or more, then the beam would have to be retargeted to each of the enormous number of aimpoints associated with a large-scale Soviet attack on _Minuteman_. In addition, the beam would have to be pointed precisely (0.05 microradians or less) during the illumination period. Therefore, the satellite beam weapon system would easily be saturated in a large Soviet attack. The accelerator associated with a space-based neutral beam system could be quite large -- probably many tens of meters. Research programs in

---

* A substantial part of the neutral beam research in the U.S. and U.S.S.R. is in direct support of the controlled thermonuclear reactor program. The neutral beam would be in the "fuel" for a system which would contain the plasma with a magnetic field.

progress in the U.S. suggest that the technology required to develop a small accelerator system may be available in the future. However, it is not impossible to imagine 100,000 kg launch weight payloads for a single beam weapon satellite. Assuming a 20,000 kg launch capability for the shuttle for a 1000 km orbit, over five shuttle loads would be required for each station. Moreover, the beam must not look too close to the earth or the atmosphere will strip away an electron and the beam will be deflected in the earth's magnetic field. A reasonable minimum altitude would be about 200-300 kilometers so that Soviet ICBMs or SLBMs on depressed trajectories simply could not be engaged. The beam weapon statellite will also require an auxiliary sensor to locate the RVs. If low orbit satellites equipped with long wavelength infrared sensors are used, they are vulnerable to direct attack and could be susceptible to simple midcourse penetration aids. A space-based radar would also be susceptible to the same well known threats, particularly electronic countermeasures. Thus, even if the very difficult scientific problems are solved, the technological and engineering problems associated with a space-based neutral beam weapon are formidable ... to say nothing of the costs. Thus, it can be predicted with reasonable confidence that the neutral particle beam research in progress will not lead to a space- based weapon system in the foreseeable future.

However, proponents emphasize that the U.S. should continue research on the possible use of neutral hydrogen beams as weapons with emphasis on the production of intense, sharply collimated hydrogen atom beams and the design of smaller accelerators. The objective of this research would not specifically be oriented toward weapon system development, but rather to prevent "technological surprise," particularly since the U.S.S.R. is reported to have an active program in this field.

The enormous technical problems associated with the neutral particle beam and its restricted use to RV destruction in the exosphere, with no capability to destroy the booster during the very vulnerable boost phase, have led many analysts to the conclusion that the most promising DEWs for use in a space-based BMD are high-power lasers.

SPACE-BASED LASER WEAPON REQUIREMENTS FOR BALLISTIC MISSILE DEFENSE

The central issues associated with the space-based BMD system are the size of the deployed system (and, therefore, the costs) and its technical credibility to absorb a specified Soviet threat. It is apparent that the BMD mission of greatest interest is "damage denial" for the U.S. in a large-scale Soviet ballistic missile attack. "Large-scale" has come to mean a near simultaneous launch of at least 1000 sea-launched ballistic missiles (SLBMs) and land-based intercontinental ballistic missiles (ICBMs) (perhaps as many as 1500 with 750 in reserve for NATO, France, China, etc.). To put the impact of such an attack in perspective, the attack would involve about 6000-6300 one-megaton equivalent warheads, most of which would be from the ICBMs. The majority would be ICBMs since the principal long range targets for these expensive systems would be located in the U.S. The ICBMs could destroy all hardened military targets in the U.S. (including ICBM silos), SAC bases, and essentially all U.S. urban/industrial areas. Only 500 of these warheads could destroy the top 100 U.S. cities and place at risk the lives of 125 million U.S. citizens. In fact, it can be shown that about 2800 warheads could destroy all U.S. military installations that could be used for retaliation (and deter the attack in the first place), in addition to destroying the U.S. as a viable twentieth century society. The Soviet military establishment could easily mount this attack with their currently deployed ballistic missile systems.[3] Such an attack would be within the parameters of their "warfighting" strategy, and would not be constrained by the very high warhead limits allowed under SALT II.[4] The ability to destroy the boosters from space before the warheads could be deployed by the upper stage (the MIRV "bus" which deploys and independently targets each RV) would mean a major shift in the strategic balance and a dramatic move away from a policy of "mutual assured destruction" toward one of "mutual assured survival."[5]

The number of laser "battle stations" required for the BMD mission is simply the product of the number required for full-earth coverage and the number required to be in the potential battlespace at all times. The number of satellites required for full-earth coverage has been determined by Luders.[6] Typically 60 to 40 satellites would be required, for 550 to 1000 km polar orbits while at 4000 km the number of satellites required decreases to about ten. Complete coverage from geostationary orbit (40,000 km) would, of course, require only three satellites. The number of lasers required to be in the "battlespace" at all times depends on the "vulnerability" or "lethality" criterion of the target and the capability of the laser to deliver a lethal dose of energy in the shortest time possible. Senator Malcom Wallop (R., WY),[7] a proponent of a space-based laser BMD, suggests that "several common household light bulbs illuminating each square centimeter" of a presumably aluminum booster skin for several seconds would result in the destruction of the booster. If the "light bulb" is assumed to emit an energy flux of 100 watts, then the illumination time is 10 seconds, then the incident lethal "fluence" would be about 1000 joules/cm^2.

The Soviet booster must be illuminated by this lethal fluence before the MIRV bus starts to deploy the RVs. That is, the booster must be destroyed before MIRV bus deployment in order to ensure that the space-based laser will not be required to attack each RV separately. It is generally accepted that the laser weapon cannot illuminate the reentry vehicle with a lethal engergy dose of infrared energy during its midcourse and reentry trajectory. That is, the RVs are very "hard" to laser weapons because they are designed to survive the launch phase, midcourse and reentry thermal environments (primarily the latter) and successfully detonate and destroy even hard targets. Thus, the booster must be destroyed between the time it is above the clouds and atmosphere, (which would preclude the propagation of an intense collimated laser beam) and the initiation of the deployment of RVs by the MIRV "bus." The access time to destroy the booster is, therefore, about 100 seconds less than the time a typical ICBM would require to complete the launch phase prior to MIRV/RV deployment. This means that the "access time" for the space-based laser to destroy the attacking booster would be about 200 seconds. If Senator Wallop is correct and the typical "lethality" level of present Soviet boosters is 1000 joules/cm^2, then the booster would be destroyed in one second with an incident power flux of one kilowatt per square centimeter. Thus, the kill rate would be one per second and assuming that the same length of time would be required to retarget the laser to another booster, each laser could destroy 100 boosters in the 200 second "access time." Thus, about 10 laser "battle stations" would have to be in the "battlespace" at all times. The size of the deployed space-based BMD would simply be the product of the number of lasers required in the battlespace and the number required for full-earth coverage, i.e., about 100 battle stations for a 1000 booster attack assuming that each laser can deliver the lethal flux of 1000 kilowatt per square centimeter for one second from a

4000 km orbit. Senator Wallop[8] suggests that a system of 24 laser "battle stations" in 800 nautical mile orbits could cope with a Soviet launch of 1000 missiles. Each laser would be a 5 megawatt (MW) hydrogen flouride chemical laser emitting at a wavelength of 2.7 micrometers and the energy focused on the target with a 4 meter (M) mirror. This system is characterized as a "5/4" system. Senator Wallop[9] indicated that each laser would carry enough fuel for the laser (hydrogen and flourine) to have "1000 shots, meaning each could cope with the theoretical contingency of a thousand missiles launched beneath it in an almost simultaneous launch." Obviously this comment refers to laser capacity and not to the limitations imposed by the "access" time to the booster and the time required to illuminate the booster with a lethal dose and retarget time. His comments on laser system effectiveness appear to relate to an attack of a few hundred ICBMs launched from the Soviet Union, e.g., "as few as ten (battle stations) could fend off a small attack, such as 300 SS-18s."[10] This seems to be supported by a recent report in <u>Aviation Week</u>[11] on the effectiveness of the 24 battle station "5/4" system. That article indicates that for a simultaneous launch, a few tens of missiles can be destroyed, and if the launch occurs over fifteen minutes (a long time), 300 Soviet boosters would be destroyed. This could account for as many as 4000 warheads[12] under the optimistic assumption that all of the SS-18s are destroyed.

The size of the deployment of a "5/4" system in 1200 km orbits to negate a near simultaneous 1000 Soviet booster attack is very large. As indicated previously, about 10 lasers would have to be in the battlespace to accommodate a 1000 booster attack using the optimistic assumption of 1000 j/cm^2 booster "lethality" and a laser system capable of delivering that fluence in 1 sec. The "5/4" system can deliver less than 1/10th of the energy required from a 1200 km orbit (actually closer to 1/20th) in 1 sec. Therefore, the irradiation times must be at least 10 seconds which would mean 100 lasers in the battlespace at all times. At a 1200 km orbit altitude the number of satellites required for full-earth coverage is 30.[13] Consequently, the deployment actually involves about 3000 laser "battle stations", simply prohibitive from anyone's perspective. Alternately, a small deployment would involve a much more capable laser, e.g., perhaps a 15 meter, 15 megawatt device. However, it is clearly not necessary to have all-earth coverage at all times. For example, a space-based BMD to counter only land-based ICBMs need only cover the north latitudes. Also the U.S. and NATO presumably have some idea where Soviet ballistic missile submarines are deployed and only those portions of the ocean's surface need to be continuously observed. With these restrictions a 24 battle station space-based BMD could be quite effective.

It is now generally agreed that the "5/4" space-based laser system, the laser weapon technology which will be demonstrated in the late 1980s, would have limited potential for a significant BMD capability to counter the near simultaneous launch of 1000 Soviet missiles. However, the "5/4" system could have a significant capability against lesser threat levels, a significant anti-satellite (ASAT) capability, and could provide the technological stepping-stone to a space-based BMD, the major objectives of the U.S. laser program.

THE VULNERABILITY OF BALLISTIC MISSILES IN THE BOOST PHASE AND LASER WEAPON COUNTERMEASURES

Several recent publications have suggested that Senator Wallop's estimate that ballistic missile boosters are vulnerable to laser radiation equal to "several household light bulbs per square centimeter" may be quite low. The casings of all solid propellant boosters in the U.S. inventory are fabricated from glass-reinforced plastics (*Minuteman*, *Polaris*, *Poseidon*, and *Trident*). Several parts of the MX, including the MIRV bus, will be fabricated using the new Dupont polyimide resin, Kevlar. These materials are similar to the materials used to fabricate reentry vehicles and in that framework are called "ablation" materials. (The term "ablation" is taken from the meteor physicists[14] and describes the absorption of heat by meteors through surface vaporization and decomposition, i.e., by surface mass removal.) These RV heat shield materials obviously protect the reentry vehicle in the severe thermal environment associated with atmosphere reentry, i.e., the reentry heating problem was solved in the U.S. almost twenty-five years ago. These materials also include carbon-like composites which are much more effective thermal protection systems (higher "heat of ablation") than glass-reinforced plastics. To put the performance of these ablation materials in perspective, the maximum turbulent heat transfer rate to a reentry vehicle during reentry occurs at about 50-60 kft altitude (15-18 km) and is about 10,000 BTU/ft^2/sec, expressed in the English units commonly used in engineering design.[15] Expressed in metric units, this would correspond to an energy flux density of 11 kilowatts per square centimeter, ten times greater than the figure used above. U.S. solid propellant boosters, therefore, could be quite "hard" to laser radiation, particularly when compared to the postulated metallic Soviet booster. Future Soviet boosters could be that hard or harder. It should also be emphasized that the illuminated spot must remain in a fixed position during the illumination period, i.e., it is assumed that the booster does not spin. Henderson[16] points out that "even slow rotation of the target could reduce the effective (illumination time) to milliseconds and thereby increase the required laser brightness by at least ten."

The possibility that Soviet boosters could be very hard has recently been discussed in various publications. Aviation Week and Space Technology,[17] a strong proponent of laser weapons, suggests that Soviet booster systems of the future could be "hardened to levels of 10-20 kilojoules per square centimeter." The same article goes on to observe that "because of postulated levels of hardness that a laser countermeasures program might provide for the U.S.S.R., the Defense Department emphasizes the use of 25 megawatt, 15 meter diameter lasers in large numbers." Similarly, Henderson[18] suggests that "state-of-the-art booster hardness of a few tens of kilojoules per square centimeter" should be considered in determining the "range and laser system parameters to size a HEL (high energy laser) system." The impact of a 20 kilojoule per square centimeter hard booster on the size of a space-based laser BMD system and/or the level of the device technology required could be significant. That is, the number of battle stations required or the capability of the laser device would have to be increased.

However, space-based laser weapon proponents have advanced the reasonable arguments that while countermeasures to laser weapons in the boost phase are technically feasible, the weight penalty to provide adequate protection may be prohibitive for the Soviets. First, one must consider how thermosetting resins or carbon-like ablation materials behave when exposed to a large heat flux. The surface of these materials pyrolyses rapidly and forms a layer of carbon. The surface temperature rises rapidly and in much less than 1 sec. reaches values as high as 3500° Kelvin to 4000° Kelvin. The surface radiates energy in proportion to the fourth power of the temperature (Planck's law) and at 4000°K the energy radiated is 1.5 kilowatts per square centimeter. If the incident flux were 1 to 1.5 kilowatts per square centimeter, the surface would be in radiation equilibrium, i.e., the surface would reradiate all of the incident radiation continuously. Therefore, the booster would be protected even if illuminated by the laser weapon indefinitely. Thus, the incident flux of laser energy must be in excess of a kilowatt per square centimeter to have any effect at all on these ablation materials. Also, the weight penalty to provide a 10 kilojoule per square centimeter capability may not be prohibitive, at least for the Soviets. The "heat of ablation" of even a modestly effective ablation material is about 2000 BTUs per pound[19] and 5000 BTUs per pound or more for a high performance material. Thus the weight penalty for a 100 foot long, five foot diameter missile would be about 3700-9300 pounds.

THE AUXILIARY SENSOR SYSTEM FOR A SPACE-BASED LASER BMD: NO SMALL TECHNICAL PROBLEM

In the continuing debate regarding the ability of high-power lasers to provide an effective space-based ballistic missile defense, there has been very little discussion of the auxiliary sensor system which the laser weapon requires. The laser beam is, of course, very narrow and cannot be used to acquire and track the Soviet boosters, barring the near-term development of an agile beam laser similar to the phased-array radar. A colocated passive infrared system or an active phased-array radar would be required. The radar option would appear to be less desirable than a passive infrared system because of the radar's vulnerability to ground-based and missile-borne electronic countermeasures. The passive infrared sensor would certainly be able to detect and track the enormous radiation from the rocket's plume, (50-100 kilowatts in the near infrared (2-5 micrometers)) produced by hot water vapor, carbon dioxide and solid particulates. The laser beam must be given the exact position of the booster structure above the plume. The infrared sensor, therefore, must be able to locate cooler missile skin with the enormous plume radiation as background. The cooler booster is presumably a "gray" or "black" body with a maximum close to 10 micrometers, so that a second infrared system is required in addition to the plume tracking system. Also, the booster may not be uniformly vulnerable. That is, the region near the fuel tanks of liquid boosters may be the positions of maximum vulnerability rather than the areas near the nozzles, etc. The inference is that the auxiliary sensor system may be required to obtain a fairly high quality infrared image of the booster in the presence of the rocket plume, identify the area to be illuminated to less than one-half a meter in diameter and pass this information over to the pointing and tracking system. The acquisition of a high quality infrared image of a 100 foot target in a radiation background of many tens of kilowatts at ranges of 3000 to 6000 kilometers is a formidable application of present or projected infrared sensor technology. The "jitter" requirement for the pointing and tracking system of 0.05 microradians is also a difficult technical requirement. The major point is that the auxiliary sensor system, while not impossible to perceive in the rapidly changing field of infrared imaging sensors, does present a major technical challenge.

NEW LASER DEVICES: THE FUTURE HOPE FOR SPACE-BASED LASER BMD AND THE WAVELENGTH DISPUTE

The present space-based U.S. laser program is reportedly divided into three parts:[20]

1. The "Alpha" program to build a 2-3 megawatt hydrogen flouride chemical laser to "demonstrate the feasibility of extrapolating this technology to the 5-10 megawatt level in the 1980s;

2. The Large Optics Demonstration Experiment (LODE) to demonstrate a "4 meter diameter primary mirror and associated beam control for experimental use" in the 1980s; and,

3. Talon Gold to demonstrate "an advanced acquisition, tracking and precision pointing system scheduled for testing with the space shuttle in FY 1987 to track targets up to 1500 kilometers with an accuracy of 0.2 microradians, scalable to 0.1 microradians."

These stated program objectives are obviously modest from anyone's perspective relative to a ballistic missile defense capability. However, this three pronged space-based laser program has two objectives: 1) the development over the next ten years of a "5/4" system with a significant ASAT capability; and, 2) to provide the basis for a technological stepping-stone to a ballistic missile defense capability.

Recently there has been a clear indication that the high-power laser's DoD program managers have come to the belief that short wavelength, UV lasers would be a more promising avenue of research to provide a significant space-based BMD capability. Dr. Robert S. Cooper, the Director of the Defense Advanced Research Project Agency (DARPA) has suggested that "high energy laser research should be oriented away from the emphasis on long wavelength chemical lasers and replaced by accelerated research on short wavelength lasers".[21] The development of short wavelength, high-power ultraviolet (UV) lasers would, indeed, be dramatic. The size of the laser power required would be less than 5 megawatts and the size of the mirror required would be less than 5 meters. Also, even hardened boosters are probably more vulnerable to pulsed UV radiation than to continuous wave infrared radiation. These "excimer" devices produce laser radiation in the ultraviolet by passing an electric discharge through a noble gas halide, e.g., xenon flouride. However, as Henderson[22] points out, these laboratory devices are extremely inefficient, typically converting 1 to 3 percent of the system's prime power to UV laser radiation. This implies a need for a 400 megawatt prime power source, a prohibitive size and weight for space applications. Thus, a major breakthrough will be required to make this technology useful for systems application.

Henderson[23] points out "that there are five basic types of high energy lasers for potential weapon system applications: combustion-driven (IR) gas lasers, electrically excited (UV) gas lasers, chemical (IR) lasers, free electron lasers and solid state, neodymium glass or neodymium YAG lasers. Of the five types, only the chemical laser is now being actively pursued as the laser device for a space-based weapon." The free-electron laser is a new contender for weapon application, and was first demonstrated by Madley in 1975.[24] In this device, a relativistic beam of electrons is passed through an alternating magnetic field. Coherent radiation is extracted with a very narrow line width and the device is tunable from the ultraviolet to the infrared region. Recent experiments at the Lawrence Livermore Laboratory of the University of California suggest that these devices could produce very high laser power efficiently.

SPACE-BASED LASER WEAPONS: WHAT TYPE OF PROGRAM SHOULD THE U.S. HAVE?

It is fairly clear that the deployment of a space-based laser BMD represents a formidable application of what is clearly immature technology at present. In addition there is the possible Soviet development of methods for the defeat of such a system. Also, it is becoming increasingly clear that the main thrust of the U.S. laser weapon program may not be in the correct wavelength region for the BMD mission. Thus, there are many analysts who would criticize the basic direction of the program. Proponents argue that it is impossible to predict accurately the future course of technology. For example, there were many capable scientists and engineers who believed the intercontinental ballistic missile would not work or that the warheads would never survive the reentry thermal environment. Then there is the possibly overriding consideration of the possibility of "technological surprise" by the Soviet Union. The Soviet Union apparently has a large program in directed energy weapons and has a very aggressive manned and unmanned military space program. It is mandatory, therefore, that the U.S. have an adequate R&D program in laser weapons in order to be in a position at all times to understand clearly the technical value of Soviet developments and to be able to respond to any Soviet initiative in directed energy weapons. The principal program policy issue, therefore, is the relative size of the laser weapons program and the expected payoff. The total laser weapon budget is about $250-$300M, according to Aviation Week and Space Technology. The total military space budget is $2.2B[25] This includes a variety of Air Force and Navy communications satellites ($550M), the Global Positioning System (GPS-$277M), the defense meteorological program (DSMP-$294M), the anti-satellite program (ASAT-$213M), the military space shuttle facilities ($582M) and launch support ($275M) programs. The Air Force RDT&E budget also contains elements relevant to the general military space program,[26] e.g., $120M - Defense Support Program Early Warning Satellites, ($40M) -- Space Surveillance and "Special Activities" ($1.1B). Thus, the budget for the procurement of new "space assets" of direct military value is $1.6B, after removing the $528M for the space shuttle launch facilities. The $1.1B "special activities" presumably are devoted entirely to space-based reconnaissance assets and as such do not represent new military space capability. In this framework, the laser weapon budget is relatively large, 15 to 17 percent of the military space budget. At this level of funding a policy-maker could make the argument that the payoff for this investment over the next decade, much less a generation, should be significant. Certainly, a doubling of the laser budget (as some proponents suggest) would mean that it would equal 35 percent of the military space budget and should require a significant payoff over the next decade. There are those who would argue that the military space program is too small at 0.8 percent of the DoD budget. One analyst[27] suggests that the U.S. military space program should be expanded significantly to include a high quality MIRV bus/RV tracking and early warning/intelligence satellites, an extensive ocean surveillance satellite system and a satellite space surveillance system. As indicated earlier, a sophisticated boost-phase tracking satellite system eventually could provide a space-based BMD with a "conventional" non-nuclear interceptor located in the U.S.[28] A policy-maker should consider a space-based laser program, therefore, in the framework of the correct allocation of resources available to the military space program for the greatest possible payoff to the U.S. in terms of expoiting space for national security purposes over the next ten to twenty years. Those purposes must include a firm requirement for the avoidance of "technological surprise" in new beam weapons by the Soviet Union. The initiation of over a dozen programs in nuclear delivery systems in the early 1960's, many of which were later cancelled, shows that incorrect allocation of defense resources could have a significant negative effect on the national security position of the United States. Clearly it is premature to judge the ultimate utility of high energy lasers in a space-based BMD system. Thus, the proper allocation of resources between laser weapons and other possible space-based assets so as to maximize the exploitation of space by the U.S. for national security purposes is one of the key defense issues of the 1980s.

SUMMARY AND CONCLUSION

In summary, while technical issues concerning space-based laser BMD remain; the system to be demonstrated in the late 1980s may well provide significant effectiveness for ASAT missions, and limited but important BMD coverage.

NOTES

1. "High Frontier", The Heritage Foundation, Washington, D.C., 1982.

2. Luders, R.D., "Journal of the American Rocket Society", February 1961, p. 179.

3. "The Military Balance: 1981-82", International Institute for Strategic Studies, London, England, p. 105.

4. Friel, P.J., "U.S. and Soviet Strategic Technologies and Nuclear Warfighting: A Comparison", MacMillan Press., Ltd., London, 1982, p. 98.

5. "High Frontier".

6. Friel, "U.S. and Soviet Strategic Technologies and Nuclear Warfighting", p. 98.

7. Wallop, Senator M., "Opportunities and Imperatives for Ballistic Missile Defense", Strategic Review, Fall 1979, p. 13.

8. Ibid.

9. Ibid.

10. Ibid.

11. Aviation Week and Space Technology, April 12, 1982, p. 18.

12. Ibid. p. 18.

13. Luders, R.D., "Journal of the American Rocket Society", February, 1961, p. 179.

14. Opik, "Physics of Meteor Flight in the Atmosphere", Interscience Publishers, New York, 1958, p. 60.

15. Martin, J.J., "Atmospheric Reentry", Prentice Hall, Englewood, N.J., 1966, p. 99.

16. Henderson, "Space-Based Lasers:", p. 42.

17. Aviation Week and Space Technology, p. 18.

18. Henderson, "Space-Based Lasers:", p. 42.

19. Martin, "Atmospheric Reentry", p. 99.

20. Aviation Week and Space Technology, p. 18.

21. Ibid.

22. Henderson, "Space-Based Lasers:", p. 42.

23. Aviation Week and Space Technology, p. 18.

24. Madley, J.M.J., et. al., "First Operation of a Free Electron Laser", Physics Review Letters, 1977, Vol. 38, p. 892.

25. Aviation Week and Space Technology, March 8, 1982, p. 15.

26. Martin, J.J., "Atmospheric Reentry", Prentice Hall, Englewood, N.J., 1966, p. 99.

27. Friel, P. J., "New Directions for the U.S. Military and Civilian Space Programs", International Security Dimensions of Space, Fletcher School of Law & Diplomacy, Tufts University, Eleventh Annual Conference on International Security Studies, April 1982, to be published.

28. Ibid.

3 Implications of Arms Control Agreements and Negotiations for Space-Based BMD Lasers

Alan M. Jones, Jr.

INTRODUCTION

This chapter examines the implications of existing arms control agreements and prospective negotiations for the development, testing, and deployment of space-based BMD lasers. The major portion of this chapter is devoted to an analysis of the provisions of the ABM Treaty relating to space-based and laser BMD systems and components and of the meaning of related terms used in the Treaty. This part of the study also examines the potential relevance of near-term BMD systems (*Sentry* and the HOE concept) for space-based BMD lasers. In addition, it assesses prospects for the 1982 ABM Treaty review conference.* The remainder of the chapter addresses other existing and potential arms control agreements relevant to space-based BMD lasers, notably the Outer Space Treaty, but also a Soviet proposal to prohibit new types of weapons of mass destruction. Finally, the chapter examines Soviet and unofficial U.S. proposals to limit or prohibit space weapons.

IMPLICATIONS OF THE ABM TREATY

The ABM Treaty prohibits space-based lasers developed and tested for ballistic missile defense.** Article V, paragraph 1, of the Treaty prohibits all mobile ABM systems and components, specifically including spaced-based systems. It states:

---

* This chapter was completed in August 1982, prior to the review conference.

** Except with reference to the ABM Treaty, the terms "ABM" and "BMD" are used interchangeably in this chapter.

> Each Party undertakes not to develop, test, or deploy ABM systems or components which are sea-based, air-based, space-based, or mobile land-based.[1]

This provision appears to be categorical and all-encompassing, and that appears to have been the intent of the U.S. and Soviet negotiators who drafted its language at SALT I. However, the provision also raises a series of questions about the terms employed. For example:

(1) What is "development" and how does it relate to testing?

(2) What are "ABM systems or components"? and

(3) What is "space-based"?

In addition, the provision raises questions about those types of activities related to space-based BMD lasers that would be permitted under the ABM Treaty. This chapter addresses such questions.

THE MEANING OF "DEVELOPMENT" IN THE ABM TREATY

The FY 1982 Arms Control Impact Statement (ACIS) for Ballistic Missile Defense provides an agreed interagency definition of "development" in the ABM Treaty.[2] Its text is worth citing in full.

The meaning of the term "development," as used in the Treaty, is as follows:

> The obligation not to develop such systems, devices or warheads would be applicable only to that stage of development which follows laboratory development and testing. The prohibitions on development contained in the ABM Treaty would start at that part of the development process where field testing is initiated on either a prototype or breadboard model.

This statement, taken from the 1972 Senate testimony of Gerard Smith, the Chief U.S. Negotiator at SALT I and the then Director of the U.S. Arms Control and Disarmament Agency (ACDA), was added to the FY 1982 ACIS to clarify the meaning of "development" and the consequent restrictions on R&D programs for BMD systems, notably land-mobile systems and components for the Low Altitude Defense System (LoADS) associated with some proposals for MX ICBM deployment.* Because the FY 1982 ACIS was sent to the Congress in mid-January 1981, it represented an interagency consensus at the end of the Carter Administration. However, there is no indication that this issue is being reopened by the Reagan Administration, at least in the ACIS arena.

The statement by Smith is important because it came as a written response to a question by Senator Henry Jackson (D., Wash.) on the meaning of "development" in Article V of the ABM Treaty (and on the use of that term in the Treaty as a whole). In making the response, Smith was not speaking on his own or for his agency, but for the Nixon Administration on this issue.

However, Smith's statement is not the only Nixon Administration declaration on the meaning of "development" in the ABM Treaty or on the Treaty's implications for space-based BMD lasers. In addition, a variety of Defense Department officials made detailed statements on the issue, usually under questioning related to R&D on laser BMD system. Those statements basically corroborated Smith's position but are worth reviewing because they indicate the state of Administration thinking on the meaning of the then newly-concluded treaty and the range of consensus (and, in some cases, ambiguous impressions) on the issue. In addition, the

---

* The LoADS system was initially intended to support the Multiple Protective Shelter (MPS) basing mode for MX favored by the Carter Administration. This mode was cancelled by President Reagan in October 1981, and it appears (as of August 1982) that initial deployment of MX will be in a Closely--Spaced Basing (CSB) mode. Unlike MPS, CSB deployment of MX would be confined to a small area (about 10 square miles) and would not require mobile BMD components for terminal defense. In one view, MX in CSB could be adequately defended by fixed BMD components consistent with ABM Treaty deployment limits (i.e., 100 interceptors and launchers, 18 smaller missile site radars and two larger perimeter acquisition radars). In another view, the <u>Sentry</u> BMD system being studied for possible deployment with CSB could consist either of fixed components (permitted by the ABM Treaty) or of mobile radars or interceptors. For example, one report states that CSB planning envisages "deployment of a ballistic missile defense system using a mobile radar."[3] In the latter case, development of mobile <u>Sentry</u> BDM components would be subject to the same restrictions as those envisaged for the LoADS system.

exchanges also indicate the breadth of interest in 1972 among defense-oriented Senators in space-based laser technology for BMD missions. It is noteworthy that all these exchanges (including Smith's statement) occurred in hearings before the Senate Armed Services Committee, which was investigating the Treaty's impact (and that of the Interim Agreement limiting strategic offensive forces) on DoD budget items. No similar questions on this issue were raised in the comparatively brief official ratification hearings on the SALT I agreements before the Senate Foreign Relations Committee.

Because of the importance of the definition of "development" as it applies to space-based BMD lasers, it is useful to treat the various statements of Defense officials at some length. It is also useful to note that their remarks occurred not in their prepared statements, which typically restated the Treaty provisions without elaboration, but in response to questioning.

The question of space-based systems arose early in testimony by then Secretary of Defense Melvin Laird. Senator Peter Dominick (R., Colorado) asked whether either SALT I agreement would "impede satellite-based counterforce system developments." In a written response, Laird stated that, while the Interim Agreement contained no such restriction, "the ABM Treaty specifically prohibits space-based ABM system."[4] (Laird also noted that the 1967 Outer Space Treaty "explicitly prohibits the placing in orbit of weapons of mass destruction." The possible implications of his Treaty for space-based BMD lasers are discussed later in this chapter.)

The issue of the specific impact of the Treaty on space-based lasers was raised by Senator Barry Goldwater (R., Arizona) and addressed by Laird in the following written exchange.[5]

> Question: For my money, we should have long since moved on the space-based systems with boosting-phase destruction with shot, nuces [*sic*], or lasers. I have seen nothing in SALT that prevents development to proceed in that direction. Am I correct?
>
> Answer: With reference to development of a boost-phase intercept capability of laser, there is no specific provision in the ABM Treaty which prohibits development of such systems.

> There is, however, a prohibition on the development, testing, or deployment of ABM systems which are space-based, as well as sea-based, air-based, or mobile land-based. The U.S. side understands this prohibition not to apply to basic and advanced research and exploratory development of technology which could be associated with such systems, or their components.
>
> There are no restrictions on the development of lasers for fixed, land-based ABM systems. The sides have agreed, however, that deployment of such systems which would be capable of substituting for current ABM radars, shall be subject to discussion in accordance with Article XIII (Standing Consultative Commission) and agreement in accordance with Article XIV (amendments to the Treaty).
>
> Secretary Laird. Could I add, in answer to Senator Goldwater's question, it does have an effect on the defensive systems, but not on the offensive systems.
> (The information follows:)
>
> No. Space-based ABM systems are prohibited by Article V of the ABM Treaty which states in part:
>
> "Each party undertakes not to develop, test, or deploy ABM systems or components which are sea-based, space-based, or mobile land-based."

This exchange underscored both the Treaty's prohibition on space-based ABM systems of any type and the Treaty's less stringent limits on fixed, land-based lasers for ABM missions. (The latter limits, set forth in Agreed Statement D, are discussed later in this chapter.)

The specific question of whether "research" as well as "development" was prohibited by Article V, and thus of what constituted development, was repeatedly raised by Senator Jackson. In questioning Dr. John Foster, Jr., then Director of Defense Research and Engineering, Jackson stated that Article V of the Treaty "raises a real question here whether you can actually engage in research." Foster did not answer at the time but instead made the following written insertion into the record. Note that this statement was made before the Smith quotation cited at the outset of this section and that it set the theme for all subsequent Nixon Administration statements on this subject. It draws a very sharp distinction between permitted laboratory development and prohibited field (i.e., post-laboratory) development and testing.[6]

> Article V prohibits the development and testing of ABM systems or components that are sea-based, air-based, space-based, or mobile land-based. Constraints imposed by the phrase "development and testing" would be applicable only to that portion of the "advanced development stage" following laboratory testing, i.e., that stage which is verifiable by national means. Therefore, a prohibition on development - the Russian word is "creation" - would begin only at the stage where laboratory testing ended on ABM components, on either a prototype or breadboard model.

A similar exchange occurred when Jackson asked Ambassador Smith about the meaning of "development". Even more than Foster, Smith refused to be drawn into an oral discussion and instead chose a written insert. Because Smith's response forms the basis of the current official interpretation of "development", the exchange is worth quoting in full.[7]

> Senator Jackson. In Article V of the Treaty, the term "develop" appears in connection with an undertaking not to "develop, test or deploy" certain ABM systems or components. How do you define the term "develop" as it is used in Article V?
>
> Mr. Smith. We have given a good deal of analysis to this question; and it is a technical question and I would ask you if I could not submit to you in writing our interpretation of what that term means. I think we can do that very quickly.
>
> Senator Jackson. Will you also take into account the further question as to whether there is any other meaning of the term "develop" associated with its use in other provisions of the accords?
>
> Mr. Smith. Yes, sir.
> (The information follows:)

> Article V of the ABM Treaty and an Agreed Interpretive Statement (F) obligate the U.S. and U.S.S.R. "not to develop, test or deploy" mobile ABM systems, rapid reload devices, or ABM interceptor missiles for delivery of more than one independently guided warhead.
>
> The SALT negotiating history clearly supports the following interpretation. The obligation not to develop such systems, devices or warheads would be applicable only to that stage of development which follows laboratory development and testing. The prohibitions on development contained in the ABM Treaty would start at the part of the development process where field testing is initiated on either a prototype or breadboard model. It was understood by both sides that the prohibition on "development" applies to activities involved after a component moves from the laboratory development and testing stage to the field testing stage, wherever performed. The fact that early stages of the development process, such as laboratory testing, would pose problems for verification by national technical means is an important consideration in reaching this definition. Exchanges with the Soviet delegation made clear that this definition is also the Soviet interpretation of the term "development".
>
> Consequently, there is adequate basis for the interpretation that development as used in Article V of the ABM Treaty and as applied to the budget categories in the DoD RDT&E program places no constraints on research and on those aspects of exploratory and advanced development which precede field testing. Engineering development would clearly be prohibited.
>
> No. The term "development" appearing in Article IV of the Treaty is used in the same sense as the term "develop" in Article V.

The most prolonged discussion of this subject occurred late in the hearings and involved Army Lt. General W.P. Leber, the *Safeguard* ABM System Manager, General Bruce Palmer, the acting Army Chief of Staff, and General John D. Ryan, the Air Force Chief of Staff. Jackson established the point with both Leber and Ryan that the only way that the U.S. could monitor

Soviet laser development and testing "in a ABM mode", if such testing occurred in the laboratory, would be by on-site inspection. Jackson noted that no national technical means of verification existed to monitor testing, but he asked "how do we monitor development work in a laboratory?" General Ryan replied that "as you well know, there is no way we could."[8]

Jackson then raised the question, "what do you mean by 'develop'?" (referring to the Article V prohibition).[9] Leber responded with a general statement that again stressed the field-testing component of development.[10]

> General Leber. My point is that development is a very major undertaking; it is not something that you can complete in a laboratory. True, some of the first work with radars and some of the first work with components of the interceptor went on in laboratories but early in their development they started testing interceptors, they started testing components of radars, putting them on the air. Those are the kinds of things that you can observe and detect.
>
> In the case of our site defense program, we are now in the early stages of a prototype demonstration where we are going to develop a system and test it out in the Pacific before we know whether it would be a system we could deploy. This is what development means to me, not just the beginning work in the laboratory but the whole program, leading to a decision on whether or not you have a system.

The discussion was broadened into generic meanings of "development" as applied to weapons systems. Chairman John Stennis (D., Miss.) offered the view that "development" did not refer to basic research but to testing and deployment. General Palmer stated his view that "development" referred to "the total creation of a system" and that the term covered the entire process "from beginning to end." Leber added that the term "can include some work in the laboratory which could not be detected but other work which could be necessary to develop fully an ABM system could be detected." This series of exchanges clouded the sharp laboratory-field testing distinction that other officials had drawn, and Palmer submitted a written insert on the meaning of the Article V prohibition, which repeated the Administration's previous theme.[11]

> Based on further research with regard to Article V of the ABM Treaty and the implications arising from the undertaking not to "develop, test, and deploy" certain ABM systems, I have since been advised by my Staff that there are some understandings which evolved during the negotiations in the use of these terms. I also understand that a more complete answer to a similar question raised yesterday in the Committee's hearings with the negotiating team members is being provided. The obligation undertaken by Article V is applicable only to that stage of development which follows laboratory development and testing. Research in these areas is therefore permitted. Exchanges with the Soviet delegation, I am told, made clear that this definition was also the Soviet interpretation of the term "development".

Perhaps the most notable aspect of these exchanges is the degree to which the Nixon Administration relied on written inserts into the testimony of both DoD and ACDA officials to set forth a unified and coherent view on the meaning of "development" in the ABM Treaty and specifically as the prohibition on space-based BMD systems and components in Article V related to lasers. There are probably several reasons for this. One that became clear in the hearings is that the discussion of the meaning of "development" in Article V became entangled with the broader issue of the development and testing of fixed, land-based ABM systems and components "based on other physical principles," which the Treaty permitted. Another reason is that the meaning of "development" in the ABM Treaty differed from the common understanding of that term within the U.S. defense community. As the statements of Stennis and Leber illustrate, that community regarded "development" as a continuous process leading from research to procurement. In contrast, the ABM Treaty established an artificial distinction in which the main criterion was not the weapons acquisition process but rather the ability of each side to monitor the other's weapons programs by its own intelligence collection assets.

In sum, the effective meaning of the prohibition on "development" of mobile ABM systems or components (including space-based systems or components) in Article V of the ABM Treaty is that it bans activities that meet all of the following three requirements: (1) they are conducted in the field (i.e., where they are monitorable by intelligence collection systems through such means as imagery, radar, and telemetry collectors); (2) they are clearly ABM-related (i.e., they are "tested in a ABM mode" against " strategic ballistic missiles or their elements in flight trajectory"); and (3) they are uniquely associated with mobile ABM systems. The first of these criteria underscores the 1972 Administration statements that prohibited actions would, in effect, begin with field testing of a prototype model.

The second criterion emphasizes that the prohibited activities are only those that are directly (and detectably) associated with ABM missions and functions. This point was established in the 1972 hearings when Senator Strom Thurmond (R., S. Carolina) asked if the ABM Treaty would prohibit the U.S. from "developing an ABM laser type weapon, for instance on tanks?" Secretary Laird gave a partially correct answer: "No, it would not apply to weapons that did not have an ABM capability".[12] Laird apparently misunderstood Thurmond's reference to an ABM laser on tanks, which the Treaty would prohibit as a mobile ABM system. In addition, Laird's reference to "ABM capability" needs to be interpreted in terms of Article VI (b) of the Treaty, under which the sides agree "not to give missiles, launchers, and radars, other than ABM interceptor missiles, ABM launchers, or ABM radars, capabilities to counter strategic ballistic missiles or their elements in flight trajectory, and not to test them in an ABM mode." Although the Treaty limits the capabilities of non-ABM radars (in Agreed Statement F), it does not limit capabilities of other non-ABM components. The U.S. unilateral statement on the meaning of "tested in an ABM mode" states that, in the U.S. view, the treaty is not intended "to prevent testing of non-ABM components for non-ABM purposes" ("U.S. Statement B. Tested in an ABM Mode".)[13] The Treaty does not restrict the development, testing, or deployment of components (other than phased-array radars) that are not tested or deployed for ABM purposes. The Treaty would not restrict tank-mounted lasers for air defense (not ABM) purposes (to modify Thurmond's example to fit Laird's answer). It also would not restrict development, testing and deployment of space-based lasers as anti-satellite (ASATs), so long as they were not "tested in an ABM mode".

The third criterion emphasizes that the prohibition applies only to mobile (including space-based) ABM components. The provision does not apply to ABM components that are developed and tested in a fixed, land-based mode and that would be compatible with the types of ABM systems and basing deployments specified in Article III of the Treaty, but that could also be deployed in a mobile basing mode (so long as those ABM components were not tested or deployed in a mobile mode). The importance of this distinction is illustrated in development and testing of the Low-Altitude Defense System (LoADS) mobile components that could be used in the *Sentry* BMD system.* That system would provide terminal defense of MX ICBMs. Testing and deployment of *Sentry* components in a mobile mode would clearly require revision of the ABM Treaty to modify or eliminate the prohibition on mobile land-based ABM systems or components. The issue of *Sentry* compatibility with the Treaty would arise only when field "testing in an ABM mode" began of components that were clearly (i.e., detectably) designed to be mobile, for example, testing of a mobile *Sentry* radar against "strategic ballistic missiles or their elements in flight trajectory." Although such testing would be required as part of full-scale engineering development, substantial testing of *Sentry* components could be conducted in a fixed mode as part of the development cycle.

The applicability of these three criteria for space-based BMD lasers is as follows. First, laboratory development and testing of laser BMD components would be permitted, because such activities could not be monitored by national technical means. Second, activities involving space-based laser components that were not solely intended for BMD or had not been "tested in a ABM mode" would be permitted under the Treaty. For example, laser pointing-tracking devices could be launched into orbit and used to track satellites, even though those same devices could also be used as part of a space-based BMD laser system (e.g., to reduce beam jitter). Third, (as discussed in the following section) development and testing of fixed, land-based BMD lasers could be conducted that would have some relevance to space-based BMD laser systems (and which could occur before space-based BMD laser prototypes were available for testing). This experience could be valuable despite the differences between beam propagation in the atmosphere and in space and between intercept of incoming RVs and of missiles during boost-phase.

---

* As noted above, the LoADS development program has been replaced by the *Sentry* BMD program. The issue of compatibility with the ABM Treaty would be the same for both LoADS and *Sentry* provided that the *Sentry* system employed one or more mobile BMD components.

THE MEANING OF "ABM SYSTEMS BASED ON OTHER PHYSICAL PRINCIPLES"

In addition to limiting existing types of ABM systems consisting of radars, interceptor missiles, and launchers of interceptors, the ABM Treaty also provides for restrictions on "ABM systems based on other physical principles and including components capable of substituting for ABM interceptor missiles, ABM launchers, or ABM radars." This restriction is set forth in Agreed Statement D* to the ABM Treaty, the text of which is reprinted below.[14]

> In order to insure fulfillment of the obligation not to deploy ABM systems and their components except as provided in Article III of the Treaty, the Parties agree that in the event ABM systems based on other physical principles and including components capable of substituting for ABM interceptor missiles, ABM launchers, or ABM radars are created in the future, specific limitations on such systems and their components would be subject to discussion in accordance with Article XIII and agreement in accordance with Article XIV of the Treaty.

The implications of this agreed statement (especially as it applied to lasers) received considerable attention in the 1972 Senate Armed Services Committee hearings on the ABM Treaty. This matter was raised early in the hearings, when Secretary Laird responded to the questions from Senator Goldwater cited above. The portions of that answer that apply to the agreed interpretation are worth repeating here.[15]

> With reference to development of boost-phase intercept capability or lasers, there is no specific provision in the ABM Treaty which prohibits development of such systems.... There are not restrictions on the development of lasers for fixed, land-based ABM systems. The sides have agreed, however, that deployment of such systems which would be capable of substituting for current ABM components, that is, ABM launchers, ABM

---

* This statement was originally designated as Agreed Interpretation E to the Treaty.

> interceptor missiles, or ABM radars, shall be subject to discussion in accordance with Article XII (Standing Consultative Commission) and agreement in accordance with Article XIV (amendments to the Treaty).

What is notable in Laird's statement is that it specifically addresses the development and deployment of lasers for boost-phase intercept. The Laird statement indicates that, in contrast to the prohibition on development of space-based ABM systems, the prohibition on ABM systems "based on other physical principles" applies only to the deployment, not to the development and testing, of fixed, land-based BMD lasers. Although the ABM Treaty's restrictions on geographical deployment of systems would preclude a situation in which fixed land-based BMD lasers could operationally intercept ICBMs or SLBMs during boost-phase, there is no prohibition on testing of such lasers against strategic ballistic missiles during boost-phase, even though such testing would be operationally relevant only to space-based lasers. Although such testing probably would be of limited applicability to space-based intercept (because of differences in beam propagation), the option for such testing exists within the terms of the Treaty and might be useful in analysis of such factors as beam direction and dwell time.

The interpretation of Agreed Statement D as permitting development and testing of future BMD systems was repeated in the State Department's analysis of the Treaty. The State Department description sets the interpretation in the context of other Treaty provisions.[16]

> A potential problem dealt with by the Treaty is that which would be created if an ABM system were developed in the future which did not consist of interceptor missiles, launchers and radars. The Treaty would not permit the deployment of such a system or of components thereof capable of substituting for ABM interceptor missiles, launchers, or radars: Article II (1) defines an ABM system in terms of its function as a "system to counter strategic ballistic missiles or their elements in flight trajectory", that such systems "currently" consist of ABM interceptor missiles, ABM launchers and ABM radars. Article II contains a prohibition on the deployment of ABM systems or their components except as specified therein, and

> it permits deployment only of ABM interceptor missiles, ABM launchers, and ABM radars. Devices other than ABM interceptor missiles, ABM launchers, or ABM radars could be used as adjuncts on an ABM system provided that such devices were not capable of substituting for one or more of these components.

Although the agreed statement would prohibit deployment of systems capable of substituting for ABM components, it would not restrict deployment of devices to supplement existing ABM components. This distinction was established during testimony by then DDR&E Foster. In response to a question from Senator Margaret Chase Smith (R., Maine), Foster stated that "nothing in the agreements ... prevents" continued research and development of lasers and that the agreement would not affect laser R&D. Foster reaffirmed: "The agreement does forbid the replacement of the currently allowed defense, that is interceptor missiles, by a laser system. ... Its deployment would be prohibited by the Treaty." However, Foster went on to add that lasers could be deployed as auxiliary devices.

> A laser could be used as part of an auxiliary designator system but it could not be used in substitution for the prime detector, that is, the ABM radar, or interceptor missile component.[17]

Foster repeated the distinction between permitted development and testing and prohibited deployment of fixed, land-based laser BMD systems in the following exchange with Senator Jackson. This exchange illustrates the potential for confusion between the obligations in Article V regarding mobile systems and those in Agreed Statement D regarding future fixed systems.[18]

> Dr. Foster. One cannot deploy a fixed land-based laser ABM system which is capable of substituting for an ABM radar, ABM launcher, or ABM interceptor missile.
>
> Senator Jackson. You can't even test; you can't develop.
>
> Dr. Foster. You can develop and test up to the deployment phase of future ABM system

> components which are fixed and land-based. My understanding is you can develop and test but you cannot deploy. You can use lasers in connection with our present land-based Safeguard system provided that such lasers augment, or are in addendum to current ABM components. Or in other words, you could use lasers as an auxiliary piece of equipment but not as one of the prime components either as a radar or as an interceptor to destroy the vehicle.

Ambassador Smith also emphasized that fixed laser R&D for BMD purposes would be permitted and that only deployment of such systems was barred. In his opening statement, Smith stated that "the parties have agreed that no future types of ABM systems based on different physical principles from present technology can be deployed unless the Treaty is amended."

(In his memoirs, Smith describes the process by which the agreed statement came about and also its interpretation. As to the latter, Smith states:

> Thus, taking the agreed understanding together with Article III, systems employing possible future types of components to perform the functions of launchers, interceptors and radars are banned unless the Treaty is amended. This statement was initialed by Semenov and me on the day the SALT agreements were signed. As an initialed common understanding, it is as binding as the text of the ABM Treaty.)[19]

In response to questioning, Ambassador Smith repeatedly denied that this understanding (or the Treaty as a whole) would constrain development of laser BMD systems. Asked by Senator Smith if the Treaty would "affect development of a laser ABM system by the United States," Ambassador Smith replied that "development work, research, is not prohibited, but deployment of systems using those new principles would not be permitted unless both sides agree by amending the Treaty."[20] A similar question (by Senator Goldwater) stated: "Under this agreement are we and the Soviets precluded from the development of the laser as an ABM?" Ambassador Smith replied: "No, sir."[21] A question (by Senator Jackson) on the Treaty's impact on Soviet technology development drew the following response from Smith:

> This Treaty...placed no limitation on the technology, the development of technology of

> radars, launchers, and interceptors. It does place limits on the technology of systems using other physical principles, but it would be fair to say that it does not limit them to the present state-of-the-art.[22]

The Treaty's effect on BMD development was raised again in testimony by the Joint Chiefs of Staff. Senator Goldwater repeated Ambassador Smith's statement that, under the Treaty, the U.S. was "allowed to continue with laser research and development." He asked if that was General Ryan's (Air Force Chief of Staff) understanding. Ryan replied: "Yes, sir." Asked the same question, General Palmer (acting Army Chief of Staff) initially stated that laser ABM R&D was "basically prohibited." However, he later amended his statement as follows: "There is no limit or understanding of a limit on R&D, in the futuristic systems, but [the understanding - sic] would require an amendment of the Treaty or further agreement to deploy such a system."[23]

These statements led to the following further exchange which again emphasized the difference in treaty obligations between laser BMD development and deployment.[24]

> Senator Goldwater. Then what you are saying, if the Army or any of our research and development agencies suddenly came along with a breakthrough that would enable us to get the power to develop the optical mechanism, would it mean that we couldn't deploy the anti-ballistic missile capability?
>
> General Ryan. That's correct.
>
> General Palmer. That's correct.
>
> Senator Goldwater. Do you both believe that?
>
> General Palmer. Without further agreement.
>
> Senator Goldwater. Again, Mr. Chairman, I think this is indicative of the great misunderstanding, and this one point alone is going to keep me from voting for the SALT until I find out what we can do. I don't like the idea that what has been the world's great technological nation, which it no

> longer is, can't develop something of its own inventive necessity where we can have a literally foolproof anti-ballistic missile system; I don't think it is around the corner; I don't think it is near. The men in the blue suits have been talking about it.
>
> General Ryan. The development, when you speak of the development, is not banned. The development --
>
> Senator Goldwater. Bad?
>
> General Ryan (continuing). Is not banned, such as Hardsite defense. The development--
>
> Senator Goldwater. But the deployment?
>
> General Ryan. The deployment of the systems is banned.

The summation of this discussion was made by Lt. General Leber, the Army's ABM system manager.[25]

> The only limitation in the Treaty, and it is in the ABM Treaty; it is not in the Interim Offensive Agreement at all, is that either side, the Soviets or the United States, would not use a laser device to substitute for any other component part of the ABM system. You could use laser technique to improve any of your existing components -- radar interceptor -- those are the main components, but if you propose to substitute, for example, a laser device for the interceptor, that would be prohibited, an amendment to the Treaty would be required for deployment.
>
> That is a very narrow area now that we are talking about; it has nothing to do with ICBMs, nothing to do with the defense systems in general. The only restriction is that you would not substitute a laser device for one of the components of your ABM system.

Finally, General Palmer reiterated that, unlike space systems, the Treaty "does not prohibit the development in the fixed, land-based ABM system. We can look at futuristic systems as long as they are fixed and land-based." General Palmer added: "The [Joint] Chiefs [of Staff] were aware of that and had agreed to that and that was a fundamental part of the final agreement."[26]

Unlike the extensive discussions just cited, the Senate Foreign Relations Committee hearings contained only one brief reference to the Treaty's limitations on laser BMD systems. In response to questioning by Senator George Aiken (R., Vermont), then Secretary of State William D. Rogers stated: "Under the agreement, we provide that exotic ABM systems may not be deployed and that would include, of course, ABM systems based on the laser principle." Ambassador Smith elaborated this point as follows:

> Unless the Treaty is amended, both sides can only deploy launchers and interceptors and radars. There are no inhibitions on modernizing this type of technology except that it cannot be deployed in mobile land-based or space-based or sea-based or air-based configurations. But the laser concern was considered and both sides have agreed that they will not deploy future type ABM technology unless the Treaty is amended.[27]

The administration statements at that time of the ratification hearings remain the U.S. position on development and deployment of BMD systems "based on other physical principles." The FY 1982 Arms Control Impact Statement (ACIS) on Ballistic Missile Defense (in contrast to the FY 1981 statement) reaffirms the distinction between permitted development and testing and prohibited deployment of such systems. The relevant portion of the ACIS is reprinted below. Although this statement represents the views of the Carter Administration, it is unlikely (in view of the statement's consistency with the 1972 testimony) that this position will be substantively revised by the Reagan Administration. While the Treaty allows

> development and testing of fixed, land-based ABM systems or components based on other physical principles such as lasers or particle beams and including such fixed, land-based components capable of substituting for ABM interceptor missiles, ABM launchers, or ABM radars, such systems or components may not be deployed under the terms of Article III and an agreed statement in connection with Article III, unless specific limitations on such systems and

their components are discussed and agreement is reached to amend the Treaty.[28]

The main purpose of presenting this evidence in such extensive detail is to establish clearly what the agreed statement on "other physical principles" does and does not permit with regard to laser BMD systems and components. A second purpose is to establish the relationship between this understanding and the prohibition on space-based BMD systems and components in Article V. The record of official testimony in 1972, and recent statements, support the conclusion that the ABM Treaty does not (and was not intended to) limit laser BMD research and development. It is also clear from these statements that, in this context, permitted "research and development" is understood to include the full range of RDT&E leading up to operational deployment. The ABM Treaty distinguishes between BMD systems deployed at operational complexes (limited in Article III) and BMD systems used for development and testing and located at test ranges (addressed in Article IV).* Article IV specifically states: "The limitations [on operational deployment] provided for in Article III shall not apply to ABM systems or their components used for development and testing."[29]

The Treaty thus permits, for example, the construction of one or more full prototype laser BMD systems, with no limitations on power, mirror size, etc., or on the use of ABM radars in conjunction with the laser for pointing and tracking. The understanding also permits the testing of such a laser "in an ABM mode" against "strategic ballistic missiles or their elements in flight trajectory." There is also no geographical restriction on the location of such prototypes, except that they be at test ranges. A common understanding to the Treaty designates those ABM test ranges that existed in 1972, but Article IV provides that the exemption for ABM systems used for R&D applies both to those locations and to "additionally agreed test ranges." Therefore, there is nothing in the Treaty to prevent the U.S. from locating one or more BMD laser test ranges at sites where conditions would be optimal for BMD laser testing (as long as those sites are within the U.S.).

---

* "Operational deployment" is used here to mean deployment of BMD systems for the defense of the national capital area or of ICBM silos, in accordance with the geographical limitations set forth in Article III of the ABM Treaty and modified by the 1974 Protocol.

The relationship between the agreed statement on "other physical principles" and the undertaking in Article V on "space-based" BMD systems or components can be simply stated. The agreed statement permits, for fixed, land-based BMD lasers, most of those activities that Article V prohibits for space-based BMD lasers. Because neither side possessed such lasers in 1972, no limits were set on their characteristics (analogous to the power-aperature limits on phased-array radars). Instead, the agreed statement states that "specific limitations" would be discussed following full-scale testing but prior to deployment. Moreover, there is no restriction on the kind of testing done. It could be for boost-phase intercept as well as for terminal defense against RVs, even though boost-phase intercept would apply operationally only in space. There is also no clearly-stated restriction on the number of tests that could be conducted prior to discussion or agreement on limitations. In short, any testing "in an ABM mode" of fixed, land-based lasers that would be relevant to space-based BMD lasers can be conducted within the terms of the Treaty. (Later sections will examine what restrictions might exist on space-based laser development, testing and deployment that would be relevant to space-based BMD lasers but would not involve "testing in an ABM mode.")

THE MEANING OF "SPACE-BASED" IN THE ABM TREATY

The discrepancy between permitted development and testing of fixed, land-based BMD lasers and the prohibition of development and testing of space-based BMD lasers raises the subject of the interpretation of "space-based" in the ABM Treaty. This matter would need to be addressed within the context of non-laser exoatmospheric BMD systems, such as the Homing Overlay Experiment (HOE) concept being studied by the Army. This system would be intended for mid-course intercept of ICBMs and SLBMs and, in most descriptions, would involve the use of space-borne long-wave infrared (LWIR) sensors on probes and of homing intercept devices that would kill by direct impact and would be guided to their targets by additional LWIR sensors. The LWIR sensors would discriminate RVs from decoys and penetration aids on the basis of heat and mass and would be impervious to "spoofing" by countermeasures designed to defeat ABM radars. BMD defenses that employed both an LWIR-impact kill overlay system and a radar-interceptor terminal defense ("underlay") system could be much more effective than either system alone.[30]

The HOE concept is relevant to consideration of the ABM Treaty's implications for space-based BMD lasers because the viability of treaty's prohibition on "space-based" ABM systems or components is likely to be affected sooner by such an overlay system than by BMD laser development.*

It appears unlikely that the U.S. will take an official position on the issue of the HOE concept's compatibility with the Treaty's ban on space-based BMD systems and components until the concept and its supporting technology are further developed. At this preliminary state, however, there appears to be a spectrum of possible view on the space-based aspect of HOE, in response to the current concept. One view is that HOE would be permitted under the Treaty because its components would normally be deployed in fixed, land-based launchers and would be launched into space only during testing or use. In this sense, the system would be analogous to ICBMs in silo launchers, which are also only launched into space during testing and use and which would not be regarded as "space-based." (The system would also be analogous to anti-satellite systems that were launched from ground sites rather than deployed in space.)

Another view is that the current HOE concept would result in a "space-based" system, and, therefore, its development, testing, and deployment would be prohibited by the Treaty. In this view, the fact that the system could operate only in a space environment and that its components (e.g., the LWIR probes) would need to remain in space for the duration of their mission would make them "space-based." In this

---

* The HOE concept also involves the use of multiple-warhead BMD interceptors, which are prohibited under Article V of the ABM Treaty as supplemented by Agreed Statement E.[31] Because this limitation would not apply to space-based BMD lasers, the miltiple-warhead aspect of the HOE concept is not examined here. However, it should be noted that, although the prohibition on such components is in a separate paragraph of Article V than the prohibition on "space-based" BMD systems or components, support for the HOE concept could lead to efforts to amend the ABM Treaty to remove Article V in its entirety, including the undertaking with regard to "space-based" systems. Such efforts might not be significantly influenced by the resolution of the question of "whether or not an overlay system such as HOE would be "space-based" as that term is used in the ABM Treaty.

view, a HOE system would differ fundamentally from ICBMs which merely move through space and do not perform their basic mission in space. (However, there would still be an analogy with ASATs launched into space to conduct attacks.)

A third view is that the compatibility of a HOE system with the ABM Treaty cannot be determined *a priori* but would need to be discussed and agreed between the sides at the time that HOE "development" (i.e., field testing) began. In this view, such discussion and agreement would not necessarily require an amendment to the Treaty or explicit common understanding in the negotiating record. It would be sufficient if one side stated that a specific overlay system was compatible with the Treaty and if the other side did not contest that statement. (However, the other side would then have the right to deploy a comparable system.) In this view, the ABM Treaty, as a bilateral agreement, can be interpreted to mean whatever the sides agree that they want it to mean, either by amendment or by informal understanding.

The relevance of the HOE example and the variety of interpretations of its compatibility with the prohibition on "space-based" systems arises in part because of the absence of an agreed definition of that term in the ABM Treaty and its supporting understandings. As in the case of other ambiguous terms in the Treaty (notably "strategic ballistic missile," discussed briefly below), the absence of a precise definition of "space-based" reflects a substantive disagreement between the sides. This matter was not discussed in the ABM Treaty ratification hearings but did arise in the Senate Foreign Relations Committee's hearings on the 1967 Outer Space Treaty. As official statements at those hearings indicate, the U.S. sought to reserve its position on the issue of definition of outer space and the boundary between "national space" (i.e., air space) and outer space. Then U.S. Ambassador Arthur Goldberg stated that the U.S. favored further study of the definition of outer space and the ratification of the Outer Space Treaty would not prejudice the U.S. position. Then Secretary of State Dean Rusk presented the main U.S. rationale for deferring a general delimitation of outer space.[32]

> Secretary Rusk. There seem to be few actual problems whose actual determination depends on the limits of outer space and soverign space.
>
> For example, I think it is generally accepted that anything that is in orbit is

in outer space. But, if we were to try to find a precise definition, there might be at this point still some very complex issues in knowing where to draw that ceiling, that dividing line, between outer space and national space, and it might indeed raise questions about the ability to enter outer space and come out of it again.

We do think this is a matter that requires further study, but at the present time we do not see practical problems that turn on this question that are of international importance and we want to study further where the balance of our own interests would lie if we get to a point where there is any inclination to put a specific number of miles as dividing line between national space and outer space.

Senator McCarthy. Mr. Chairman, could I ask about a distinction? I believe you talked about sovereign space and outer space.

Secretary Rusk. National space and outer space.

Senator McCarthy. Then it would become sovereign space in the sense you assert jurisdiction over it.

Secretary Rusk. In that sense it has been traditional law over the centuries in that sovereignty runs from the center of the earth --....

Senator McCarthy. Are you going to establish something as outer space beyond the gravitational field of the earth, and is that within sovereign space?

Secretary Rusk. I think all we can say is an object in orbit is in outer space.

Senator McCarthy. If it is in orbit and subject to gravitational field of the earth.

Secretary Rusk. Yes, I would think so, sir.

The Soviets have been the main proponents of an air space/outer space boundary, which they have proposed be set at 100-110 km (or, more broadly, at 90-130 km) above sea level. Their representatives at the Legal Subcommittee of the U.N. General Assembly's Committee on the Peaceful Uses of Outer Space have argued that national sovereignty over air space might not necessarily extend to this boundary but that further talks on the regime for higher altitudes below the boundary could be conducted once a treaty had established the basic demarcation. The Soviets have also stated that, under such a treaty, space objects would maintain a right of flight over the territories of states (i.e., through national air space) on their way to or from outer space and the territory of the launching state. Such a right might be analogous to the "right of innocent passage" permitted foreign ships through states' territorial waters.

The position stated by Secretary Rusk has remained the basic U.S. view on this subject. As presented at the U.N., the U.S. position has three main arguments. First, activities in outer space have been conducted for nearly 25 years without the need for a definition or delimitation of outer space. Second, neither the Soviets nor any other country has identified a specific problem that would be solved or mitigated by the establishment of such a boundary. Third, the Soviet-proposed boundary would be artificial and would not be based on scientific or technical analysis. In addition to these arguments, the U.S. is concerned that the establishment of such a boundary could restrict outer space activities conducted by the U.S. or by friendly states. In particular, the U.S. would want to avoid a situation in which the Soviets could claim a right to attack the U.S. space shuttle in the event that an emergency required it to fly over the Soviet Union at an altitude below such a boundary.[33]

The relevance of this discussion for space-based BMD lasers is that some partial testing of lasers against ballistic missile targets might be conducted using laser platforms that were not placed into orbit. Like the components of the HOE system, they might rendezvous with a ballistic missile during the exo-atmospheric portion of its trajectory without completing a full orbit. It appears likely that only some limited types of tests might be conducted in a partial-revolution or direct ascent mode. However, the U.S.-Soviet disagreement over the definition of outer space and the resulting ambiguity of the term "space-based" in the ABM Treaty probably would permit such partial testing. Thus, the question of whether or not such testing (if feasible and desirable) would be permitted under the Treaty is moot and is not likely to be resolved until the U.S. is required by events (e.g., the prospect of HOE field testing "in an ABM mode") to address the issue and take a position.

The ambiguity as to the definition of "space-based" would have no effect on the Treaty's prohibition of full-system testing and deployment of space-based BMD lasers. Such testing and deployment would require that the laser platforms be placed in orbit, rather than launched into partial orbits. As Secretary Rusk stated, there is no disagreement that objects placed in orbit around the earth are in outer space. Rusk's statement is also consistent with the Outer Space Treaty's prohibition on deployment in space of "weapons of mass destruction" (discussed in a later section). That undertaking (in Article IV) specifically obligates States Parties to the Treaty "not to place in orbit around the Earth" such weapons. It also reflects the ambiguity of meaning of "outer space" by providing that states would not "station such weapons in outer space in any other manner" (a formulation intended in part to prohibit deep-space deployment of ICBMs). Thus, the ambiguity as to "space-based" might permit some limited testing of BMD laser systems.

MEANING OF "STRATEGIC BALLISTIC MISSILE"

Another ambiguity in the ABM Treaty with possible relevance to development and partial testing of space-based BMD lasers is the absence of an agreed definition of "strategic ballistic missile" (and the meaning of the broader concept "tested in an ABM mode"). These phases are fundamental to the definition of ABM systems and components as set forth in Article II of the Treaty. Paragraph 1 of that article reads as follows:[34]

Article II

1. For the purpose of this Treaty an ABM system is a system to counter strategic ballistic missiles or their elements in flight trajectory, currently consisting of:

   (a) ABM interceptor missiles, which are interceptor missiles constructed and deployed for an ABM role, or of a type tested in an ABM mode;

   (b) ABM launchers, which are launchers constructed and deployed for launching ABM interceptor missiles; and

(c) ABM radars, which are radars constructed and deployed for an ABM role, or of a type tested in an ABM mode.

The absence of a definition in the Treaty of "strategic ballistic missiles" was the result of fundamentally different U.S. and Soviet positions as to the characteristics of "strategic" weapons. This difference surfaced early in the SALT I talks and affected both the ABM Treaty and the Interim Agreement limiting strategic offensive arms. In the U.S. view, "strategic" weapons were those whose longer range capabilities, greater nuclear payload, and other characteristics distinguished them from "theater" or "tactical" nuclear weapons. In the Soviet view, a weapon was "strategic" if it could strike targets on the territory of the other side from its operational deployment, regardless of its other characteristics. The U.S. view was "absolute" in comparing weapons with each other, regardless of geographical considerations. The Soviet view was "relative" in emphasizing the importance of basing. U.S. acceptance of the Soviet view of "strategic" would have meant agreement that U.S. theater nuclear forces deployed in Europe and Asia (what the Soviets called "forward-based systems") would be limited under SALT and would count against U.S. ICBMs and SLBMs within the SALT ceilings. Because of this impasse, the sides essentially choose to "agree to disagree" on the meaning of "strategic" in both SALT I agreements. The U.S. also failed to obtain Soviet agreement on a definition of "tested in an ABM mode." Instead, the U.S. delegation at the talks made the following unilateral statement.[35]

B. Tested in ABM Mode

On April 7, 1972, the U.S. Delegation made the following statement:

> Article II of the Joint Text Draft uses the term "tested in ABM mode," in defining ABM components, and Article IV includes certain obligations concerning such testing. We believe that the sides should have a common understanding of this phrase. First, we would note that the testing provisions of the ABM Treaty are intended to apply to testing which occurs after the date of signature of the Treaty, and not to any testing which may have occurred in the past. Next, we would amplify the remarks we have made on

> this subject during the previous Helsinki phase by setting forth the objective which governs the U.S. view on the subject, namely, while prohibiting testing of non-ABM components for ABM purposes: not to prevent testing of ABM components, and not to prevent testing of non-ABM components for non-ABM purposes. To clarify our interpretation of "tested in an ABM mode," we note that we would consider a launcher, missile or radar to be "tested in an ABM mode" if, for example, any of the following events occur: (1) a launcher is used to launch an ABM interceptor missile, (2) an interceptor missile is flight tested against a target vehicle which has a flight trajectory with characteristics of a strategic ballistic missile flight trajectory, or is flight tested in conjunction with the test of an ABM interceptor missile of an ABM radar at the same test range, or is flight tested to an altitude inconsistent with interception of targets against which air defenses are deployed, (3) a radar makes measurements on a cooperative target vehicle of the kind referred to in item (2) above during the reentry portion of its trajectory or makes measurements in conjunction with the test of an ABM interceptor missile or an ABM radar at the same test range. Radars used for purposes such as range safety or instrumentation would be exempt from application of these criteria.

The main effect of this statement in terms of laser applications was to elaborate the phrase "strategic ballistic missile" to include any "target vehicle" whose flight trajectory resembled that of a "strategic ballistic missile." This statement reduced the potential loophole in the Treaty provided by the absence of a definition of the latter term. It meant that, in the U.S. view, the Soviets could not test BMD interceptors against, for example, IRBMs, and then claim that such testing was not "in an ABM mode" because an IRBM was not a "strategic ballistic missile." Of greater relevance to space-based lasers, the U.S. statement probably would mean that the U.S. could not test such lasers against, for example, *Scout* rockets during boost phase because those rockets would have flight trajectories "with characteristics of a ballistic missile flight trajectory." Conversely, the Soviets would be prohibited from testing space-based lasers against space launch vehicles, for the same reason. Although the 1972 statement on "testing in an ABM mode" is a U.S. unilateral statement and is not legally binding on the Soviets, there is no indication that the Soviets have a significantly different view on the statement's description of "strategic ballistic missile."

Although the statement probably would prohibit space-based laser testing against launch vehicles other than strategic ballistic missiles, the statement probably would not affect such testing against satellites that resembled ICBM or SLBM upper stages. Space objects that were specifically designed to simulate ICBMs and SLBMs in free flight would not have flight trajectories comparable to those of ballistic missiles that were launched on sub-orbital trajectories. For that reason, tests against such space objects might be unrestricted. However, there is no official U.S. view on the compatibility of such testing with the ABM Treaty, nor is any official determination likely until such testing becomes a near-term prospect. Thus, the possibility exists -- and has existed for some time -- that BMD-related testing could be conducted against satellites as a means of circumventing the ABM Treaty's prohibition on testing of space-based BMD systems or components. Although such testing would be partial, especially for space-based lasers intended to perform boost-phase intercept, it could still be of considerable relevance as part of a space-based BMD laser testing program (for example, in testing pointing and tracking, beam jitter, and deposited power levels at various distances). There are currently no agreements limiting testing or deployment of ASAT systems. (Possible future restrictions on ASATs are discussed in a later section.)

The argument presented here is that a space-based laser weapons development program would be consistent with the ABM Treaty unless a prototype of such a weapon were tested against a "strategic ballistic missile" (as defined in the U.S. statement to mean a vehicle with the same flight characteristics as a strategic ballistic missile). In this view, in the absence of an agreement limiting or prohibiting space-based laser weapons for non-BMD missions, there are no legal (vice physical) limits on the size or capabilities of space-based lasers.

A counter-argument is that a multi-megawatt, large-aperture mirror space-based laser (such as the types described in another chapter) would be "capable of substituting for" ABM components in the meaning of the agreed statement and thus would be prohibited under Article V. There are several possible reponses to this argument. One would be to assert that such a laser is not "ABM-capable" within the meaning of the Treaty. This response could be made, for example, to an inquiry in the Standing Consultative Commission (or elsewhere) concerning an "ambiguous activity" related to compliance with the ABM Treaty. A side could claim that the activity in question was not ABM-related. The other side would then need to decide whether or not to challenge such a claim and, if so, whether or not sensitive monitoring capabilities could be jeopardized and whether or not the issue was likely to be satisfactorily resolved.

Another response would be to state that the undertaking "not to give" non-ABM components "capabilities to counter strategic ballistic missiles or their elements in flight trajectory" (in Article VI (a)) applies to "missiles, launchers, or radars", not to space-based lasers and that (as discussed previously) limits on "development" of such systems would apply only if the systems were "tested in an ABM mode." It is instructive to note that "capabilities" as used in Article VI is not defined in the Treaty or (except for non-ABM radars, limited in Agreed Statement F) in the accompanying agreed or unilateral statements and that the U.S. statement related to this article refers to testing. This is a further indication that "capabilities", like "development," are, in effect, equated with field testing, if only because of monitoring requirements.)

A related counter-argument might be that very large space-based lasers would be effectively "constructed and deployed for an ABM role" (within the meaning of the definition of ABM components in Article I of the Treaty) because they were too powerful for other (e.g., ASAT) missions and would thus be prohibited under Article V. A response, besides repeating the argument that it is BMD testing of space-based lasers that the Treaty prohibits, would be to note that laser effectiveness against satellites is in part a function of range and that multi-megawatt lasers with large fuel reserves would be required to enable a side to destroy, within a brief time period, large numbers of enemy satellites at long distances from the laser platforms. The argument about excessive capability would more clearly apply to deployment of several large space-based lasers than to the launch of a single prototype for ASAT testing that would also be relevant to BMD missions. (It is not suggested here either than ASAT testing of space-based lasers could substitute for tests against strategic ballistic missiles or that either the U.S. or the Soviets would deploy space-based lasers for BMD missions without a series of tests in space against strategic ballistic missiles.)

SUMMARY OF ABM TREATY IMPLICATIONS

The foregoing discussion indicates that the ABM Treaty's prohibition on space-based BMD systems or components would nonetheless permit substantial partial development and testing of BMD lasers designed for eventual full-system testing and deployment in space.

- Article V of the Treaty prohibits field testing of prototypes of space-based BMD systems (i.e., tests in space "in an ABM mode") but would permit laboratory and other non-space testing of fixed, land-based BMD systems.

- Agreed Statement D would prohibit operational deployment of fixed, land-based BMD laser system components but would permit development and full-scale testing of prototypes at designated test ranges.

- There is no agreement on what constitutes a "space-based" BMD system that is not "placed in orbit around the Earth." This raises the possibility of testing of sub-orbital lasers against "strategic ballistic missiles or their elements in flight trajectory."

- There is no agreed definition of the phrase "strategic ballistic missiles" in the ABM Treaty. However, the 1972 U.S. unilateral statement on "testing in an ABM mode" clearly applies the term to any "target vehicle" whose flight trajectory has the same characteristics as a strategic ballistic missile flight trajectory. This would, for example, prohibit tests of orbital lasers against sounding rockets. Because the emphasis is on flight trajectory rather than the characteristics of the target vehicle, however, it probably would not preclude testing of orbital lasers against orbital targets (i.e., satellites) that were configured like booster stages of strategic ballistic missiles.

- The Treaty and associated statements do not restrict non-BMD development and testing of space-based lasers. The Treaty defines ABM system components in terms of: (1) construction and deployment as ABMs; or (2) testing "in an ABM mode." In practical terms, a device that is not tested against "strategic ballistic missiles or their elements in flight trajectory" is not a BMD system or component. This means, inter alia, that there are no restrictions on testing of space-based laser components (e.g., pointing and tracking) against non-ballistic missile targets, even though such testing would have direct BMD applications.

One means of assessing the ABM Treaty's implications for space-based BMD lasers is by analogy to the Sentry program. This analogy has added relevance because the Sentry program is currently under development and decisions about its further progress will be important for determining both the extent of activities permitted within the Treaty and the U.S. position toward modification (or abrogation) of the Treaty.

The main similarity between the mobile land-based Sentry components and space-based BMD lasers is that the Treaty prohibits development, testing, and deployment of both systems. The main differences are twofold: (1) Sentry, though mobile, would be land-based and would operate in the same environment as permitted fixed, land-based systems; and (2) Sentry would employ radars and interceptors that, although greatly improved from those of the 1972 Safeguard system, would be comparable to them and not based on "other physical principles." These differences point to an even more fundamental difference between Sentry and space-based BMD lasers: The Sentry components derived from LoADS could be operationally deployed in a fixed basing mode within the terms of Article III of the Treaty, although the Treaty's restrictions on interceptors and, especially, radars would make such a deployment of questionable value as a means of protecting ICBM silos. (Either the earlier LoADS concept of one radar to 1-2 interceptors would need to be modified or, if concept were maintained in the fixed sites, the interceptors deployed would be constrained by the radars to a small percentage of the permitted total.) In contrast, laser BMD components could not be operationally deployed within the terms of the Treaty and the prototypes that could be constructed at one or more test ranges probably would not be directly convertible into components suitable for deployment in outer space.

The relevance of the Sentry analogy is that BMD testing of both Sentry and laser components can be conducted at test ranges, so long as such testing is in a fixed basing mode. All BMD testing of both Sentry and lasers that does not require the use of mobile platforms could be carried out within the terms of the Treaty. Conversely, mobile platforms designed for Sentry components and for space-based lasers could be developed and tested provided that such platforms were not directly coupled with Sentry or laser components that had been tested against "strategic ballistic missiles or their elements in flight trajectory." This testing of platforms could be accomplished using simulated components. Put differently, the Treaty does not clearly prohibit the launching into outer space of an orbital prototype of a laser BMD system or component, so long as that prototype is not tested in an ABM mode and so long as the prototype is distinguishably different from a fixed, land-based BMD laser prototype that had been tested "in an ABM mode."

The fact that Sentry would employ radars and interceptors and that a laser BMD system would employ "components capable of substituting for ABM interceptor missiles, ABM launchers, or ABM radars" does not affect the relevance of the Sentry analogy for development and testing. It would (as noted above) negate the relevance of Sentry in terms of operational deployment. However, as the testimony in the 1972 hearings and recent U.S. statements make clear, the Treaty permits as full a range of BMD-related development and testing for fixed land-based "exotics" such as lasers as it does for "conventional" BMD systems such as Sentry.

The limitation of the Sentry analogy comes not so much in what activities are permitted under the Treaty as in the applicability of those activities for planned deployment modes. In the case of Sentry, the applicability is clear and direct. There is a relatively small amount of testing of mobile Sentry components that could not be done from fixed sites. Such testing probably would be mainly "proof" tests of the system's compatibility with a mobile basing mode, including its ability to function following repeated movement. However, the Sentry components would not be mobile when in use, and the integration of those components could be accomplished without mobility. In short, the difference between a permitted and a prohibited Sentry system is essentially one in which the components were set in concrete versus one in which the same components were mounted on transporters.

In the case of space-based BMD lasers, the applicability of fixed, land-based testing would be far less direct. Beam propagation through the atmosphere would be very different than through space. Size and weight constraints imposed on space-based lasers by launch vehicle capabilities would not apply to fixed, land-based lasers. Land-based lasers could be more readily supplemented by BMD radars (for target tracking) than could space-based lasers. Because of atmospheric effects, power generation requirements would differ. Other differences could be noted as well. The point is that fixed, land-based BMD laser components probably could not be launched into space "as is" and subjected to a few "proof tests" prior to operational deployment.

As the _Sentry_ analogy suggests, the ABM Treaty constraints place a premium on the applicability of permitted development and testing. There is no Treaty obligation that all of the laser prototypes developed and tested "in an ABM mode" from fixed sites must be optimized for deployment in a fixed, land-based mode. Within the physical constraints imposed by the difference between atmospheric and space environments, some or all of those components could be made compatible (e.g., in size and weight) with deployment in outer space. Unlike the _Sentry_ example, BMD testing of fixed, land-based lasers would need to be supplemented by non-BMD testing of space-based lasers to enhance the applicability of permitted testing. ASAT testing of such lasers could be made to closely simulate tests against ballistic missile stages, and the laser components themselves could be made to resemble the characteristics (e.g., size and weight) of BMD lasers.

One view is that a significant factor in the willingness of U.S. officials fully to exploit the ABM Treaty provisions to enable partial testing of space-based BMD laser components could be possible Soviet reactions to such testing. Although Soviet responses are discussed in another chapter, it is relevant to mention two possible responses here. First, the Soviets could challenge the U.S. actions as violations of the ABM Treaty (by, for example, raising such charges in the U.S.-Soviet Standing Consultative Commission). In that event, assuming that the U.S. defended its actions, the Soviets could either drop the matter, with the tacit or explicit understanding that they would be permitted similar activities, or they might continue to assert that the U.S. was violating the Treaty, and, in the extreme event, they might withdraw from the Treaty. Second, as an alternative, the Soviets might choose not to make such a protest but instead to develop and test their own space-based BMD laser components so as to enable rapid breakout to operational deployment of such a system following abrogation of the Treaty. In this case, U.S. actions could set a precedent for actions that the Soviets might take in their space-based laser BMD program.

Although this argument has merit, there are two important qualifications that need to be added to it. First the fate of the ABM Treaty (both as a whole and in terms of its provisions relevant to space-based BMD systems) is likely to be decided on other grounds (e.g., Sentry and HOE deployment) long before the U.S. is ready to undertake activities that could be interpreted as circumventing the Treaty's prohibition on space-based BMD systems. For this reason, the issue of possible Soviet response to U.S. activities that are permitted by the Treaty and are relevant to space-based BMD lasers needs to be addressed in the broader context of the future status of the ABM Treaty. Second, some claim that the Soviets are substantially ahead of the U.S. in terms of BMD laser development, including that of space-based BMD lasers. Thus, the appropriate compliance-related issue may be the consistency of near-term Soviet laser activities with the terms of the Treaty and the U.S. reaction to those activities, rather than the Soviet response to farther-term U.S. activities.

PROSPECTS FOR THE 1982 ABM TREATY REVIEW

Article XIV, paragraph 2, of the ABM Treaty states: "Five years after the entry into force of this Treaty, and at five-year intervals thereafter, the Parties shall together conduct a review of this Treaty."[36] The Treaty entered into force on October 3, 1972. A session of the U.S.-Soviet Standing Consultative Commission (SCC) established to implement the Treaty was held to conduct the first five-year review in the Fall of 1977. A second review (almost certainly through the SCC) was required under the Treaty in the Fall of 1982.

The Treaty's provision for periodic review is common to other recent arms control agreements to which the U.S. is a party. For example, the 1968 Non-Proliferation Treaty, intended to limit the spread of nuclear weapons, provides for five-year reviews (Article VIII, paragraph 3). Similarly, the 1977 Convention on the Prohibition of Military or Any Other Hostile Use of Environmental Modification Techniques (the ENMOD Convention) also provides for quintennial reviews (Article VIII, paragraphs 1 and 2). In contrast, the Seabed Arms Control Treaty, among others, requires that an initial five-year conference be held to "review the operation of this Treaty" and to decide "whether and when an additional review conference shall be convened" (Article VII).

There are two factors, however, that separate the review provision in the ABM Treaty from those of other arms control treaties. First, the initial five-year review of the ABM Treaty was deliberately tied to the expiration of the SALT I Interim Agreement limiting strategic offensive forces. This linkage was made explicit by a U.S. unilateral statement at the time of signature of the ABM Treaty. It stated that the U.S. would regard failure to achieve satisfactory limits on strategic offensive forces to replace the Interim Agreement as constituting an "extraordinary event" (within the meaning of Article XVI of the ABM Treaty) that would justify U.S. withdrawal from the Treaty.

The second major difference between the ABM Treaty and other recent arms control agreements is the existence and functioning of the Standing Consultative Commission (SCC). The U.S. and the Soviets established procedures early in the SCC's existence that require the sides to hold two sessions per year (one in the Spring and one in the Fall) to discuss matters related to the implementation of the SALT I agreements. These sessions typically begin in March and October and last about six weeks, (however, both the timing and duration of these sessions have varied widely).

At the first ABM Treaty review conference in 1977 (convened as a special session of the SCC), the Carter Administration took the view that the SCC had, in effect, provided a continual review function. The review conference apparently was a brief series of meetings in which the sides reaffirmed their undertakings in the Treaty and expressed their overall satisfaction with its implementation.

As of this writing (August 1982), it appears very unlikely that the U.S. will propose any amendments to the ABM Treaty at the 1982 review conference. The main reason is the view that there is no pressing need for modification of the Treaty at this time. There are no fixed procedures for proposing amendments to the Treaty. Article XIV, paragraph 1, states that "Each party may propose amendments to this Treaty," and it states that entry into force of agreed amendments would follow the same procedures as those governing entry into force of the Treaty. The 1982 review conference thus provides only one of many vehicles for introducing such amendments (either within or outside the framework of the SCC), and a decision not to propose an amendment at the conference would not affect U.S. ability to make such a proposal at a later time. In particular, it would not defer U.S. proposal of amendments until the 1987 review conference.

One of the factors in the probable U.S. decision not to propose amendments at this time concerns uncertainties regarding the future status of the Sentry program. The status of Sentry, in turn, is directly (and exclusively) linked to the resolution of the MX ICBM basing issue. The Reagan Strategic Program announced in October 1981, deferred the main MX deployment decision pending completion of studies on three alternative basing modes. At that time, Secretary of Defense Caspar Weinberger stated that the Administration would make a decision among these alternatives in 1984.[37] In authorizing continued funding for the MX program, Congress subsequently directed the Administration to prepare a plan for permanent MX basing by June 30, 1983. In March 1982, the Senate Armed Services Committee voted to require the Administration to choose a permanent basing mode for MX by December 1, 1982. This date was incorporated into the Senate version of the FY 1983 DoD authorization bill, which also prohibited interim MX production or basing. Although the final bill as passed in August 1982 authorizes funds to begin MX production, the Administration has promised to decide on an MX basing mode by December. The December 1982 deadline has become the one which is operative for the Sentry program and, consequently, for any modification of the ABM Treaty to accommodate that program.

The linkage between MX deployment and possible modification of the ABM Treaty was recently reaffirmed by Dr. Fred C. Iklé, Under Secretary of Defense for Policy. In response to a question about the relationship of BMD deployment to protect MX to the ABM Treaty, Dr. Iklé stated:

> If and when there is a concrete proposal, we would obviously have to review it and see whether it would require a change in the ABM Treaty. At this time, we do not have any proposal for an ABM system to protect the MX deployment that is sufficiently advanced to make this judgment.[37]

One view is that the Administration would need to decide all of the details of Sentry deployment (the system configuration, deployment areas, numbers of radars, interceptors and launchers) in order to fashion a proposed amendment that would permit such deployment while restricting Soviet BMD options. This decision probably would not be made until 1983, or possibly later. Another view might be that the Administration would need only to decide in principle on Sentry deployment in order to propose an amendment to the Treaty to permit full-scale testing of Sentry components in a mobile basing mode. Such a decision could be made before 1983. In practice, however, it appears unlikely that the U.S. would adopt the second view, which could require a two-step approach to Treaty amendment (one to permit land-mobile testing and deployment and another to permit higher BMD component deployment levels) within a short time period. For this reason, it seems probable that the U.S. will defer consideration of ABM Treaty amendments until after the ABM Treaty Review Conference and the MX basing decision.

There is no indication that space-based BMD lasers have played or are likely to play a role in the Administration's preparations for the 1982 Treaty review. They would probably be regarded as far-future systems that would not be affected by the Treaty's restrictions for many years. In addition, to the extent that the Soviets are seen as having a relatively near-term space-based BMD laser potential, the Treaty prohibition on such systems could be viewed as a means of restricting Soviet activities. Because the ABM Treaty review is likely to be viewed in terms of short-term, narrowly-focused considerations, factors involving longer time horizons will probably be essentially ignored.

The U.S. position at the 1982 Treaty review is thus likely to involve basically two alternatives. On the one hand, the Reagan Administration could follow the Carter example and treat the review conference as essentially a "non-event" in which neither side would raise significant issues and both sides would reaffirm their intentions to fulfill their Treaty commitments. On the other hand, the Administration could decide to raise one or more issues related to Soviet compliance with the Treaty to demonstrate U.S. dissatisfaction with past Soviet activities that were viewed as counter to the spirit and the letter of the Treaty. The latter approach appears the more likely because it would depart from the 1977 practice and because it would underscore the Administration's commitment to holding the Soviets to strict standards of verification. However, any such compliance-related issues would be of the same type that could be raised by the sides during regular SCC sessions. U.S. raising of such issues at the review conference would be a means of highlighting their importance to the U.S., but this action would not otherwise distinguish the review from other SCC session. Whatever course the Administration takes in preparing for the 1982 review conference, it appears very unlikely that the U.S. will use that occasion to seek a change in the ABM Treaty.

IMPLICATIONS OF THE OUTER SPACE TREATY

The main provision of the 1967 Outer Space Treaty that has relevance for the testing and deployment of BMD lasers in space is the first paragraph of Article IV. It states:

> States Parties to the Treaty undertake not to place in orbit around the Earth any objects carrying nuclear weapons or any other kinds of weapons of mass destruction, install such weapons on celestial bodies, or station such weapons in outer space in any other manner.[39]

This provision raises the question of whether or not space-based BMD lasers would fall into the category of "any other kinds of weapons of mass destruction" and thus be prohibited by the Treaty. In other words, if the ABM Treaty restrictions on space-based BMD lasers were removed, would the Outer Space Treaty still prohibit deployment of such weapon systems? As noted above, there is no agreed definition of "outer space" in the Outer Space Treaty (or elsewhere). The U.S. position is to oppose such a formal definition, at least at this time. As also noted, however, the U.S. view was (and is) that objects deployed in Earth orbit would be in outer space, so that space-based lasers would be regarded as space objects within the meaning of the Outer Space Treaty and thus subject to its provisions.

There is no definition of "any other kinds of weapons of mass destruction" in the Outer Space Treaty. It should be noted that this phrase was formulated by the U.S. in the early 1960s. In 1962, then Deputy Secretary of Defense Roswell Gilpatric stated that it was U.S. policy not to orbit in space satellites that contained nuclear weapons "or any other kinds of weapons of mass destruction." In 1963, this same phrasing appeared in a U.S. declaration at the U.N. that later became a U.S.-Soviet joint declaration and was adapted into a U.N. General Assembly resolution. The main operative clause of that resolution contains nearly identical language to the Outer Space Treaty provision cited above.[40] Although the Soviets later proposed introducing this provision into the Treaty, the original phrase regarding weapons of mass destruction was of U.S. origin, and the U.S. agreed to the Soviet proposal of its inclusion into the Outer Space Treaty. (The 1966 U.S. Draft Treaty had dealt with the exploration of the Moon and other celestial bodies, without any arms control element.)[41]

Although the phrase is not defined in the Treaty, there are several statements of the U.S. view of "other kinds of weapons of mass destruction" in the ratification hearings before the Senate Foreign Relations Committee. These statements all indicate that the U.S. regarded "weapons of mass destruction" as weapons whose destructive effects were comparable to those of nuclear weapons. For example, under questioning by Senator J. William Fulbright (D., Arkansas), then Chairman, U.S. Ambassador Arthur Goldberg cited biological weapons.[42]

> The Chairman....What are the other weapons of mass destruction that you have in mind?
>
> Mr. Goldberg. Bacteriological, any type of weapon which could lead to the same type of catastrophe that a nuclear weapon could lead to....It does not refer to any conventional weapon. It refers to a weapon of the magnitude of a nuclear, bacteriological weapon.

Goldberg repeated the example of biological weapons in response to a question by Senator Frank Carlson (R., Kansas). Asked to define a weapon of mass destruction, Goldberg stated: "This is a weapon of comparable capability of annihilation to a nuclear weapon, bacteriological. It does not relate to a conventional weapon." Goldberg added that this provision of the Treaty would not apply to satellites that were not weapons platforms. "Observation satellites, navigation satellites, these are not covered by the Treaty."[43]

Defense Department views of "weapons of mass destruction" were similar to those stated by Goldberg and again stressed the widespread effects of such weapons. In his opening statement, then Deputy Secretary of Defense Cyrus Vance made the following characterization of such weapons.[44]

> The provision that weapons of mass destruction will not be placed in orbit around the earth undertakes to preserve man from yet another series of weapons which could threaten whole populations.

In later questioning by Senator John Sherman Cooper (R., Kentucky), Vance gave a further brief statement of this view. Asked to describe "weapons of mass destruction" other than nuclear weapons, he replied:[45]

> I believe it would include such other weapon systems as chemical and biological weapons, sir, or any weapon which might be developed in the future which would have the capability of mass destruction such as that wreaked by nuclear weapons.

The then Chairman of the Joint Chiefs, General Earle Wheeler, also testified in support of the Treaty but made no statement about the meaning of "weapons of mass destruction". Although General Wheeler was asked whether the Outer Space Treaty would affect the effectiveness of an ABM system in the atmosphere or in outer space, his reply clearly indicates that he was thinking in terms of a nuclear-armed, land-based system, not space-based lasers. He stated that the Treaty would have no impact on ABM deployment or use.[46]

No ACDA official testified at the Outer Space Treaty ratification hearings. It may be useful to note, however, that then Assistant Director of ACDA for Science and Technology, Herbert C. Scoville, Jr., states many years later that: "The deployment of laser ABMs would not be a violation of the current Outer Space Treaty banning the stationing of weapons of mass destruction in outer space."[47]

The phrase "weapons of mass destruction" has a long history in the United Nations. In 1948, the U.N. Security Council established a Commission on Conventional Armaments, which specified the scope of its activities as encompassing all weapons except "atomic weapons and weapons of mass destruction." It proposed to define the latter as follows:[48]

> [W]eapons of mass destruction should be defined to include atomic explosive weapons, radioactive material weapons, lethal chemical and biological weapons and any weapons developed in the future which have characteristics comparable in destructive effect to those of the atomic bomb or other weapons mentioned above.

This resolution is notable not only because it has remained the U.S. and Western position but also because it was passed over Soviet opposition. (The Soviet veto in the U.N. Security Council does not apply to votes on procedural matters.) Soviet refusal to accept this Western definition of the phrase did not prevent them from using the term in a wide range of arms control contexts but without a definition. In the mid-1970s, the Soviets gave the phrase a substantially broader meaning than that of the West when they introduced in the U.N. Committee on Disarmament a draft agreement on the "prohibition of the development of new types of weapons of mass destruction and of new systems of such weapons." (See the discussion of this Soviet proposal in the following section)

As discussed in detail below, despite the Soviet attempt to broaden the phrase and apply it to new technologies, the Soviets later in effect accepted the Western definition of "weapons of mass destruction." This occurred in the context of a U.S.-Soviet bilateral working group on prohibiting the development, production, and stockpiling of radiological weapons. These talks occurred under the auspices of the U.N. Committee on disarmament and led to a U.N. General Assembly draft resolution (jointly introduced by the U.S. and the Soviets in November 1979) that called on members to conclude a convention banning such weapons. In the text of that resolution, the Soviets in effect agreed to the 1948 definition of "weapons of mass destruction" as applie to nuclear, radiological, chemical, and biological weapons. This resolution constituted de facto U.S.-Soviet agreement that other types of weapons (including lasers) are not "weapons of mass destruction."

The view that U.S.-Soviet agreement exists on the definition of existing types of "weapons of mass destruction" (and that this term does not apply to lasers, whether ground- or space-based) is reflected in U.S. official statements during hearings on the unratified SALT II Treaty. In Article IX (c) of that proposed Treaty, each side "undertakes not to develop, test or deploy ... systems for placing into Earth orbit nuclear weapons or any other kind of weapons of mass destruction, including fractional orbital missiles." These statements make clear the U.S. view that, as of 1979, no U.S.-Soviet differences existed over the interpretation of "weapons of mass destruction." The statements make equally clear that high-energy lasers are not among such weapons. For this reason, they are worth quoting at some length. The exchange is between Senators John Glenn (D., Ohio) and Claiborne Pell (D., R.I.) and Mr. Thomas Graham, Jr., then ACDA General Counsel, and Ambassador Ralph Earle II, then Director of ACDA and Chief of the U.S. Delegation to the SALT II talks.[49]

> Senator Glenn. Mr. Chairman, could I ask on this, if they or we have permanently orbiting spacecraft? I presume this would

prevent any other kind of weapon being used. Do we define weapons of mass destruction?

Ambassador Earle. No.

Senator Glenn. Are those presumed as being nuclear or any weapons of mass destruction?

Mr. Graham. Senator Glenn, there is a long-standing understanding about what "weapons of mass destruction" means. That phrase appears in earlier arms control agreements. It means nuclear weapons, chemical weapons, radiological weapons and bacteriological weapons.

Senator Glenn. That was going to be the next question, chemical and bacteriological.

Mr. Graham. Yes.

Senator Glenn. They would be prohibited from being carried on board a spacecraft?

Mr. Graham. Under this provision they would be.

Senator Glenn. Is that classified as a weapon of mass destruction?

Mr. Graham. That is the generally accepted international view as to what this phrase means. It appears in a number of arms control agreements.

Senator Glenn. Have the Soviets agreed to that description?

Mr. Graham. I believe they have, yes, sir.

Senator Pell. I wonder if there might be submitted for the record an itemization of what is considered a mass destruction weapon by the United States and what are considered weapons of mass destruction by the Soviet Union.

Ambassador Earle. I think a mutuality of views exists between the two parties.

Senator Pell. List that for the record.

Mr. Graham. We will prepare that.

Senator Pell. I share Senator Glenn's curiosity. When and where did the Soviets agree to this, too?

Ambassador Earle. We will be pleased to supply that for the record, Senator.

[The following material was subsequently supplied for the record:]

The following resolution was adopted by the U.S. Convention Commission for Conventional Armaments on August 2, 1948, and supported by both the United States and the Soviet Union:

The Commission for Conventional Armaments resolves to advise the Security Council:

1. that it considers that all armaments and armed forces, except atomic weapons and weapons of mass destruction, fall within its jurisdiction and that weapons of mass destruction should be defined to include atomic explosive weapons, radioactive material weapons, lethal chemical and biological weapons, and any weapons developed in the future which have characteristics comparable in destructive effect to those of the atomic bomb or other weapons mentioned above.

2. that it proposed to proceed with its work on the basis of the above definition.

On November 2, 1979, the United States and the Soviet Union tabled a joint draft resolution in the First Committee of the United Nations on the subject of concluding an international convention prohibiting the development, production, stockpiling and use of radiological weapons. This draft resolution (U.N. Doc. A/C.1/34/L.7) has the effect of recording an endorsement by both the United States and the Soviet Union of the above-quoted U.N. definition.

There is no indication that these statements differ from the views of the current Administration. They indicate the U.S. view of the importance of the

U.S. resolution on radiological weapons as confirming U.S.-Soviet agreement on the definition of "weapons of mass destruction." There is, however, at least one important discrepancy between these statements and the record. The Soviets did not support the 1948 U.N. resolution that first defined "weapons of mass destruction" which is dated August 12, 1948.

IMPLICATIONS OF THE SOVIET PROPOSAL TO PROHIBIT NEW TYPES OF "WEAPONS OF MASS DESTRUCTION"

In 1976 and 1977, the Soviets proposed, in the U.N. Conference of the Committee on Disarmament, a draft agreement to prohibit the "development and manufacture of new types of weapons of mass destruction and new systems of such weapons." The main objective of this proposal was to go beyond the 1948 definition of "weapons of mass destruction" by providing a new definition that would add other types of weapons and ban them before they had been developed and acquired.

The main revision of the 1977 Soviet draft agreement (and the Annex to it) is reprinted below.[50]

Article I

1. Each State Party to this Agreement undertakes not to develop or manufacture new types of weapons of mass destruction or new systems of such weapons.

For purposes of this Agreement, the expression "new types and new systems of weapons of mass destruction" includes weapons which may be developed in the future, either on the basis of scientific and technological principles that are known now but that have not yet been applied severally or jointly to the development of weapons of mass destruction or on the basis of scientific and technological principles that may be discovered in the future, and which will have properties similar to or more powerful than those of known types of weapons of mass destruction in destructive and/or injuring effect.

The list of types and systems of weapons of mass destruction to be prohibited by this Agreement is contained in the Annex to the Agreement.

2. In the event that new areas of development and manufacture of weapons of mass destruction and systems of such weapons not covered by this Agreement emerge after the entry into force of the Agreement, the Parties shall conduct negotiations with a view to extending the prohibition provided for by this Agreement to cover such potential new types and systems of weapons.

3. States Parties to the Agreement may, in cases where they deem it necessary, conclude special agreements on the prohibition of particular new types and systems of weapons of mass destruction.

4. Each State Party to this Agreement undertakes not to assist, encourage or induce any other States, group of States or international organizations to engage in activities contrary to the provisions of paragraph 1 of this article.

ANNEX TO THE AGREEMENT

An approximate list of types and systems of weapons of mass destruction covered by the Agreement on the prohibition of the development and manufacture of new types of weapons of mass destruction and new systems of such weapons.

The following types and systems of weapons shall be prohibited by the Agreement on the prohibition of the development and manufacture of new types of weapons of mass destruction and new systems of such weapons:

(1) Radiological means of the non-explosive type acting with the aid of radioactive materials.
(2) Technical means of inflicting radiation injury based on the use of charged or neutral particles to affect biological targets.
(3) Infrasonic means using acoustic radiation to affect biological targets.
(4) Means using electromagnetic radiation affect biological targets.

This lists of types and systems of weapons to be prohibited may be supplemented as necessary.

In proposing an earlier version of this draft agreement, the Soviets presented working papers to the U.N. that elaborated the terms used in Article I and the Annex specifying types of weapons. In these working papers, the Soviets stated:[51]

I. New types of weapons of mass destruction

New types of weapons of mass destruction shall include types of weapons which are based on qualitatively new principles of action and whose effectiveness may be comparable with or surpass that of traditional types of weapons of mass destruction.

The term "based on qualitatively new principles of action" shall be understood to mean that the means of producing the effect or the target or nature of the effect of the weapon is new.

The term "means of producing the effect" shall be understood to mean the specific type of physical, chemical or biological action.

The term "target" shall be understood to mean the type of target, ranging from vitally important elements of the human organism to elements of man's ecological and geophysical environment, and also to networks and installations which are vitally important for human existence.

The term "nature of the effect shall be understood to mean the new type of destruction which leads either to immediate mass annihilation or to the gradual extinction of large groups of the population.

New systems of weapons of mass destruction (apart from systems which may be developed in the future when research and development of new types of weapons of mass destruction has been completed) shall include systems of weapons which assume the character of weapons of mass destruction as a result of the use of new technical elements in their strike or logistic devices.

The Soviets stated that the "means of producing the effect" included, among others, "electromagnetic waves in the optical range (light and infrared radiations)." The types of targets were three: (1) man; (2) the "human environment" (i.e., agriculture, soil, climate, upperatmosphere, and "near space," defined in terms of radiation belts; and (3) "man-made products" (i.e., "installations and networks vitally important for human existence"). The most important definition, however, was that of the "nature of the effect." The Soviets described this as follows:

> Mass annihilation of people and population, their degradation and extinction, large-scale permanent (irreversible) and temporary loss of certain human abilities, destruction and incapacitation of vitally important installations and networks.[52]

In presenting their proposal, the Soviets argued that the draft agreement would establish "an insurmountable barrier to the emergence and development of new types of weapons of mass destruction, possibly even more pernicious and devastating than nuclear weapons." It would have this effect by prohibiting types of weapons which were "either currently non-existent or were only at the stage of research and experimentation."[53] Although mentioning optical electromagnetic weapons, the Soviets laid greater stress on "acoustic sound weapons," weapons designed to damage the human reproductive system, and "ethnic weapons that attacked only certain population groups."[54]

The Soviets stated that their proposed definition of "weapons of mass destruction" was "universal in character."[55] They contrasted their approach to that of the 1948 U.N. definition, which, they claimed, failed to "make the necessary distinction between known types of weapons of mass destruction, including those that had been modernized, and new types of weapons of mass destruction which might be developed." They proposed supplementing that definition by adding a criterion related to the weapons' damaging effect on human beings.[56] The Soviets stated that the desirability of seeking specific agreements to ban specific weapons (such as radiological weapons) that might be developed did not remove the urgent need for a comprehensive agreement prohibiting "new types of weapons of mass destruction."[57]

The Western response to this position (mainly by the U.S. and U.K.) was three-fold. First, they argued that the Soviet proposal was "conceptually elusive" and lacked a "clear and generally accepted foundation" that would "ensure a harmonious relationship with existing agreements and negotiations."[58] In the U.S. view, the Soviet proposal "would merely create the illusion" of having resolved the problem and "would inevitably lead to continuous haggling over the designation of new weapons as new weapons of mass destruction."[59]

Second, the West maintained that the 1948 U.N. definition was adequate "as a comparability standard for future definition of weapons of mass destruction and as a basis for identifying new candidate weapon types as weapons of mass destruction."[60] There was no justification for amending the 1948 definition "on the grounds of new technological developments since no such developments existed."[61]

Third, the Western delegates stated that the problem of new weapons of mass destruction could best be addressed by agreement on specific limitations on specific systems. An overall agreement would be of little "practical value, since each type of weapon had its own characteristics and special set of problems."[62] The lead times for new weapons were sufficiently long to permit limits to be established as the need arose.[63] Moreover, there was a risk that the Soviet proposal would "confuse the discussion of priority items, namely, the weapons of mass destruction recognized in 1948 ... by allowing discussion of new weapons of mass destruction to overlap them."[64]

This discussion was inconclusive and was essentially subsumed in 1979 by U.S.-Soviet agreement on "major elements of a treaty prohibiting the development, production, stockpiling and use of radiological weapons" in a bilateral working group associated with the U.N. Committee on Disarmament (CD).[65] The U.N. resolution urging the CD to continue negotiation on such a treaty represented Soviet agreement, in effect, to defer further consideration of their proposed comprehensive agreement pending completion of the radiological weapon (RW) negotiations. Discussion of the two issues was essentially merged and refocused on RW. The result has been that the Soviet proposal has received little attention in recent sessions of the CD. This is indicated by the CD's 1981 report, in which the section on "new types of weapons of mass destruction and new systems of such weapons; radiological weapons," is almost wholly devoted to the latter subject.[66] The report contains a brief reference to the 1977 comprehensive proposal and calls by "some delegations" to press for such an agreement while "others felt that it would be sufficient to give periodic attention to this question." There was general support to establish a meeting of experts to review recent scientific developments and trends.[67] The Soviet proposal continues to be advocated, mainly by East European countries, but does not appear to have significant support among most CD members. Thus, the Soviet proposal appears to be essentially a non-issue for the foreseeable future.

The Soviet position on "new types of weapons of mass destruction" probably has only an indirect relationship to space-based BMD lasers. Although laser-type weapons were cited in the 1976 Soviet working paper, they did not figure prominently in the U.N. discussions and were not listed among the examples of the Annex to Article I of the 1977 Soviet draft proposal. Moreover, the primary standard applied by the Soviets in defining such weapons is massive damage to major segments of the human population, either directly or through effects on the biosphere or on structures that are "vitally important for human existence." It would be difficult to maintain credibly that space-based BMD lasers could produce such massive effects on population, environment, or vital facilities and networks. Although ICBMs and SLBMs contribute to security by enhancing deterrence, they are arguably not critical to human survival. Indeed, it would be more plausible to maintain that space-based BMD lasers would contribute to (rather than damage) human existence by destroying weapons that have long been recognized as "weapons of mass destruction." It is only if space-based lasers could be targeted against the Earth with "widespread, long-lasting, or severe effects" (to use the language of Article I of the ENMOD Convention) that they might be considered to be "new types of weapons of mass destruction."

The consistent Western position of opposition to the Soviet-proposed comprehensive agreement indicates that it is very unlikely that an arms control agreement will be concluded that limits space-based BMD lasers indirectly under the rubric of "weapons of mass destruction." Instead, the trend in U.N.-related negotiations has been and appears likely to remain under the Reagan Administration to seek agreements limiting or prohibiting specific types of weapons, such as the 1975 convention prohibiting biological weapons and the ongoing negotiations on prohibiting chemical and radiological weapons.

The CD discussions also illustrate the point that the Soviets cannot unilaterally redefine "weapons of mass destruction." The U.S. and other Western countries (all parties to the Outer Space Treaty) have indicated that they regard the 1948 definition of that term as adequate and that this definition is confined to nuclear, radiological, chemical and biological weapons. The Soviets could not alter the meaning of that term, as used in the Outer Space Treaty, to include (and thus prohibit) U.S. deployment in space of BMD lasers. If the Soviets took this view and failed to secure U.S. agreement, they could, in the extreme case, withdraw from the Treaty. They would have no right either under the Treaty or outside it to use force against such lasers (unless the U.S. had initiated conflict in space or elsewhere). In sum, so long as the U.S. adheres to its position that the term "weapons of mass destruction" (as that term is generally accepted) does not include space-based BMD lasers, there is no basis for the conclusion that the Outer Space Treaty restrictions would apply to such systems.

IMPLICATIONS OF ASAT PROPOSALS FOR SPACE-BASED LASERS

There is no agreement limiting anti-satellite (ASAT) weapons systems. In March 1977, the U.S. proposed that a U.S.-Soviet bilateral working group be established to address limits on ASAT systems (as one of a series of such groups). Soviet acceptance of this proposal led to the convening of exploratory talks on ASAT limits in June 1978 and to negotiations on such limits in 1979. The sides failed to reach agreement during these talks, and further sessions were postponed first by the Soviet invasion of Afghanistan and then by the U.S. change of administration. The Reagan Administration has yet to take a position on the matter of resuming U.S.-Soviet bilateral talks on ASAT limits or on U.S. proposals if such talks were reconvened. In general, largely because of U.S. interest in developing an ASAT system to satisfy military requirements (e.g., against Soviet surveillance satellites) and because of the difficulty of verifiably eliminating the existing Soviet ASAT capability, there appears to be little likelihood that the Administration will seek to resume bilateral talks or that it will seek other limits on ASAT systems.

THE SOVIET DRAFT TREATY PROHIBITING DEPLOYMENT IN SPACE OF "WEAPONS OF ANY KIND"

Talks on ASAT limits have recently been held in the U.N. Committee on Disarmament. These talks address a "draft treaty on the prohibition of the stationing of weapons of any kind in outer space," which the Soviets proposed to the U.N. General Assembly in August 1981. As explained in an accompanying letter from Soviet Foreign Minister Andrei Gromyko to then U.N. Secretary General Kurt Waldheim, the purpose of the Soviet proposal is to supplement the Outer Space Treaty and other agreements relating to outer space with a treaty that "precludes the possibility of the stationing in outer space of those kinds of weapons which are not covered by the definition of weapons of mass destruction."[68] (This letter is a further indication that the Soviets agree that there is a single, generally accepted definition of existing types of "weapons of mass destruction.") This draft treaty, if agreed, would clearly prohibit all space deployment of laser weapons (for whatever mission) as well as all conventional types of weapons.

The main provisions of the Soviet-proposed draft treaty are reprinted below.[69]

ARTICLE 1

1. States Parties undertake not to place in orbit around the earth objects carrying weapons of any kind, install such weapons on celestial bodies or station such weapons in outer space in any other manner, including on reusable manned space vehicles of an existing type or of other types which States Parties may develop in the future.

2. Each State Party to this treaty undertakes not to assist, encourage or induce any State, group of States or international organization to carry out activities contrary to the provisions of paragraph 1 of this article.

ARTICLE 2

States Parties shall use space objects in strict accordance with international law, including the Charter of the United Nations, in the interest of maintaining international peace and security and promoting international co-operation and mutual understanding.

ARTICLE 3

Each State Party undertakes not to destroy, damage, disturb the normal functioning or change the flight trajectory of space objects of other States Parties, if such objects were placed in orbit in strict accordance

> with article 1, paragraph 1, of this treaty.

The proposal was discussed briefly in the General Assembly's First Committee in November 1981. The Soviets offered a resolution calling on the Committee on Disarmament to add to the agenda of its Spring 1982 session an item on the "conclusion of a treaty on the prohibition of the stationing of weapons of any kind in outer space." This resolution was adopted with Western abstentions. Several Western countries (but not the U.S.) presented a resolution requesting the Committee on Disarmament (at the same session) to consider two questions: (1) that of "negotiating effective and verifiable agreements aimed at preventing an arms race in outer space;" and (2) that of "negotiating an effective and verifiable agreement to prohibit anti-satellite systems, as an important step towards the fulfillment of the objectives" of those broader agreements. This resolution was also adopted, with Soviet-bloc abstentions.[70] (The U.S. voted for the Western resolution.)

The U.S. statement in the First Committee debate provided the first statement of Reagan Administration policy on ASAT arms control.[71]

> The United States fully supports the goal of protecting outer space for peaceful purposes and is committed to avoiding a military confrontation in outer space. ...My delegation is prepared to participate fully in the discussions in the Committee on Disarmament on the question of the need for outer space arms control measures. ...[I]t smacks of hypocrisy for the Soviet Union to seek a treaty that would prohibit the stationing of weapons in outer space when in fact it is the only country that has already deployed a weapons system for destroying satellites. The existence of the Soviet ASAT system clearly complicates this entire issue. My delegation is of the view that when the Committee on Disarmament begins its discussion on the question of outer space arms control, primary emphasis should be placed on the threat posed by the Soviet ASAT system.

The Committee on Disarmament debated the Soviet proposal in April 1982. During that debate, the U.S. noted that several agreements exist (notably the Limited Test Ban Treaty, the Outer Space Treaty, and the ABM Treaty) that regulate military activities in outer space and raised the question of whether additional agreements were required. As in the November 1981 statement, the U.S. expressed its concern "regarding Soviet military activities in outer space, and in particular the development of a weapons system designed to intercept and destroy satellites." The U.S. added:

> The unilateral development of these anti-satellite weapons puts at risk the satellites of all other nations which have been launched and are used for commercial, scientific and monitoring and other lawful peaceful purposes.
>
> My government has no desire to engage in a costly arms race in outer space. The threat posed by Soviet military activities in outer space requires a prudent response on our part. At the same time, we believe the committee should continue to discuss the question of possible further arms control measures in outer space. While the United States has not yet reached a decision regarding whether further such measures would be appropriate, whether they could enhance stability, and indeed whether negotiations are feasible at this time, we are prepared to address the question with an open mind.[72]

This statement is likely to remain the basic U.S. response to the Soviet draft agreement to prohibit "weapons of any kind" in outer space.

It is clear that the Soviet proposal would prohibit deployment of space-based BMD lasers. Although the Soviet draft appears to be confined to a deployment limit, it would also have the effect of prohibiting the launching into space of lasers and other weapons that would go into orbit. This could, for example, prohibit testing of the current Soviet ASAT system in its nominal testing mode. However, it would not affect testing of a direct-ascent ASAT interceptor (such as the U.S. is reportedly developing). It probably would also not affect testing or deployment of an exoatmospheric "overlay" BMD system such as that envisaged in the HOE concept, because such systems would be sub-orbital and would not be "stationed" in space. Similarly, the proposal probably would not restrict sub-orbital or fractional-orbital testing of lasers launched into space. (The proposal would have no effect on ASAT tests of ground-based lasers, because only the target vehicles would be launched into space.)

Although no U.S. position on the substance of the Soviet proposal has been made public, it appears unlikely that the U.S. would agree to this proposal in its present form. Article 1 of the Soviet draft purports to be a simple extension of the Outer Space Treaty prohibition on space deployment of "weapons on mass destruction" to a ban that would apply to all weapons. However, the provision contains a specific reference to weapons deployed "on reusable manned space vehicles." This phrase is clearly intended to apply to the U.S. space shuttle. It seems likely that the U.S. would oppose the specific designation of the space shuttle in this provision because the shuttle is not a weapons system and would not be used for that purpose. (There would be little advantage to the U.S. in doing so because of the shuttle's vulnerability to ASAT attack and because of future U.S. dependence on a few shuttle orbiters for its entire space launch capability.) The effect of the Soviet draft language would be to set separate restrictions on the shuttle as compared to expendable launch vehicles, which are not similarly designated. This phrase probably reflects a desire to restrict shuttle operations as a major Soviet objective in seeking to limit space weapons. Even with this reference, however, the provision probably would not affect use of the shuttle as a test-bed for components (e.g., pointing and tracking devices) that could be integrated into space-based BMD laser systems, because such components (other than the lasers themselves) would not be regarded as weapons.

The most important reason for probable U.S. opposition to the Soviet draft concerns Article 3 of the proposal (Article 2 is a slightly modified restatement of Article III of the Outer Space Treaty and probably would be acceptable on that basis). Article 3 would restrict a party's obligation not to attack space objects that were (or that the party believed to be) non-weapons platforms. This formulation would permit attacks on space objects that a party believed to be armed. It would thus create a right to attack satellites that does not now exist in international law.

To oversimplify a complex subject, the use of force in international law is governed primarily by the U.N. Charter, which states recognize as setting forth fundamental international law. Article 2(4) of the Charter prohibits the threat or use of force except for the purposes of the United Nation (i.e., as directed by the Security Council, under provisions that have never been implemented due to U.S.-Soviet differences -- the authorization for U.N. forces in the Korean War being a partial exception). The sole right of states to use force independently of the U.N.

under existing international law is the right of self-defense as set forth in Article 51 of the U.N. Charter. That article provides that states have a right to immediate self-defense "against armed attack" (which the U.S. has interpreted to include collective as well as individual self-defense and as applying to "imminent" attack as well as to actual destruction). The right of self-defense does not permit the use of force by one state against the armed forces of another state unless elements of those forces are engaged in imminent or actual armed attack against the first state or its allies (and even then, in the U.S. view, the rule of proportionality of response to the scale and effects of the attack would apply).

What this brief overview indicates is that the Soviets do not have the right to attack space-based weapons (whether BMD lasers or other systems) simply because they are weapons deployed in space, any more than they have the right to attack earth-based troops, warships, or ICBMs deployed outside their territory. The Soviet proposal would create such a right. The fact that weapons deployed in space would be in violation of Article 1 of the Soviet draft would not permit the Soviets to attack those weapons. If the Soviets believed that a prohibition on space weapons was being violated, they would have the right to take a variety of actions, up to and including withdrawal from the Treaty and subsequent deployment in space of their own weapons. However, they would not have the unilateral right to "enforce" the prohibition by attacking the weapons-carrying space objects that had been deployed in violation of the agreement. (By analogy, if the Soviets deployed BMD systems in excess of ABM Treaty limits, the U.S. would have the right to abrogate the ABM Treaty. It would not have the right to attack the excess Soviet BMD deployments.)

This provision of the Soviet draft would thus result in a significant modification of fundamental international law on the use of force and would set a dangerous precedent for future arms control agreements. For this reason, and also because of the shuttle restrictions, the U.S. and other Western countries probably will oppose the Soviet draft agreement in the CD discussions. Because of this likely Western opposition, the Soviet draft proposal is not likely to be approved and consequently would have no effect on the development, testing, and deployment of space-based BMD lasers.

In addition to these specific objections, the U.S. might also seek to restrict limits on space weapons to weapons systems that currently exist. As suggested by one former participant in U.S. policy-making on ASAT arms control issues, the U.S. might use the ABM Treaty as a model for limits on space weapons.[73] As discussed earlier in this chapter, that treaty limited the deployment (and, in certain cases, the development and testing) of systems based on existing types of components. It placed no restrictions on development and testing of future types of components based on "other physical principles." By analogy, the U.S. might seek to limit "conventional" ASAT systems that rely on some type of interceptor vehicle as a kill-mechanism (whether explosive or impact) rather than "exotic" systems that employ beams of energy. Although the Soviet proposal appears intended solely as a deployment limit (despite its effects on testing of ASAT interceptors against target satellites), the U.S. could take the position that a prohibition on space-based weapons should not apply to types of weapons that do not exist. Such a position would be consistent both with the ABM Treaty analogy and also with the Western response to the Soviet proposal on prohibiting "new types of weapons of mass destruction."

If the Soviet draft were adopted in its present form or with changes that did not alter the prohibition in draft Article 1 on the placement in orbit of "weapons of any kind," such an agreement could have a significant restrictive effect on partial testing of space-based BMD lasers. As noted, partial tests that involved sub-orbital weapons (analogous to the HOE concept) probably would be permitted. Similarly, all space testing of non-weapons components would be permitted. However, testing of space-based lasers against specially instrumented target satellites configured as missile boost stages would be prohibited. This restriction would thus go beyond the limits imposed by the ABM Treaty.

OTHER POSSIBLE LIMITS ON ASAT SYSTEMS

The only other recent proposal for ASAT limits that could affect space-based BMD lasers is a Senate resolution introduced in May 1981 by Senator Larry Pressler (R., South Dakota). The Pressler resolution reads as follows:[74]

> SENATE RESOLUTION 129--RESOLUTION RELATING TO NEGOTIATIONS ON THE LIMITATION OF ANTI-SATELLITE WEAPONS SYSTEM
>
> Whereas the United States and the Union of Soviet Socialist Republics are parties to the Treaty on the Limitation of Anti-Ballistic Missile Systems, done at Moscow on May 26, 1972;

Whereas Article XII of the Treaty prohibits each party to the Treaty from interfering with the other party's national technical means of verification;

Whereas the United States and the Union of Soviet Socialist Republics increasingly depend on satellites for verification of arms control agreements, strategic warning, communications, meteorology, navigation, scientific exploration, and other missions;

Whereas the development of anti-satellite (ASAT) weapons would call into question the ability of satellites to perform these functions, threaten our freedom to operate in space, and raise the specter of a costly and destabilizing arms race in space;

Whereas in the absence of an ASAT agreement, both parties may soon complete development of new and more sophisticated ASAT weapons: Now, therefore, be it

Resolved, That is it the sense of the Senate that the President should promptly resume negotiations with the Government of the Union of Soviet Socialist Republics on a balanced and verifiable agreement limiting the development, deployment or use of weapons systems designed exclusively to intercept, damage or destroy orbiting satellites.

Sec. 2, The objective of such negotiations should be to promulgate an agreement providing for --

(1) a comprehensive non-use ban;
(2) a moratorium on further testing in space of any weapons system designed exclusively to intercept, damage or destroy orbiting satellites;
(3) dismantling and destruction of all such currently operational ASAT interceptors;
(4) stringent verification of each party's compliance with the above provisions.

Sec. 3, It is further the sense of the Senate that in pursuing agreement with the Union of Soviet Socialist Republics on ASAT

> limitations, the President should agree to no provision that would in any manner restrict the development and operation of the Space Shuttle or impede legitimate research and development (R&D) activities permitted under the SALT I ABM Treaty.

The Pressler resolution was referred to the Senate Foreign Relations Committee, where Pressler chairs the Subcommittee on Arms Control. Senator Charles Percy (R., Illinois), Chairman of the Committee, forwarded the resolution to then Secretary of State Alexander M. Haig, Jr., for comment. Secretary Haig reportedly sent a classified response that probably stated that the Reagan Administration was reviewing the U.S. policy on ASAT arms control issues. In any event, the Foreign Relations Committee took no action on the Pressler resolution, and, consequently, it was not brought to the full Senate for a vote. Thus, the resolution is a dormant issue at present, although its author could resurrect it, depending in part on the results of the CD discussions and on Administration decisions on ASAT arms control issues. Even then, however, there is no indication that this resolution would form the basis of the U.S. position on possible ASAT limits.

This resolution is clearly more ambitious than the Soviet proposal, because it would prohibit all deployments (earth- and space-based) of ASAT systems and would at least temporarily prohibit all ASAT testing. It is also more limited in scope than the Soviet proposal because it would apply only to weapons systems "exclusively designed" to attack satellites, rather than to "weapons of any kind." However, as long as the ABM Treaty remains in force, the effect of both the Pressler resolution and the Soviet proposal on space-based BMD lasers probably would be essentially the same. Both proposals would eliminate partial testing of space-based lasers against target satellites that resembled missile boost stages (and other satellite configurations relevant to BMD testing). The Pressler resolution would also prohibit testing of ground-based lasers against target satellites. To the extent that such testing was relevant to space-based BMD lasers, this would be an additional limitation. However, the resolution specifically would not affect R&D testing permitted under the ABM Treaty, including the development and testing of ground-based lasers against "strategic ballistic missiles." Thus, the resolution would have no effect on the type of ground-based laser testing that probably would be of greatest relevance to development of space-based BMD lasers.

Apart from its effects on partial testing of space-based BMD lasers, the Pressler resolution is subject to criticism on several grounds. The proposal for what Pressler called "a comprehensive ban on the actual use of ASAT weapons" is perhaps the least controversial. The intent would be to prohibit any type of attack on or interference with all space objects covered under an agreement. Such a provision would apply the prohibition on interference with U.S. and Soviet satellites used as "national technical means [NTM] of verification" of the ABM Treaty (the only agreement currently in force that prohibits such interference) to non-NTM space objects as well. However, such a proposal raises a large number of issues that would need to be addressed before agreement was reached. The more important of these issues include the following: (1) coverage of space objects under the agreement, (2) the status of space objects excluded from coverage, and (3) types of activities involving space objects that would be prohibited. A detailed discussion of these issues would be outside the scope of this chapter, but brief mention may suggest the potential complexities of these issues. First, U.S. and Soviet satellites would presumably be covered under the agreement, based on national registry in accordance with the 1975 Convention on Registration of Space Objects. However, the U.S. might also want to cover space objects registered by other countries (e.g., Intelsat V, launched and registered by France). Criteria for determining coverage of such satellites could be difficult to establish. Second, a U.S.-Soviet agreement might exclude space objects of U.S. Allies, and the U.S. would need to determine the effects of such an agreement on existing legal protections, under the U.S. Charter and the Outer Space Treaty, of space objects excluded from such a bilateral agreement. This issue could affect U.S.-Allied relations. Third, an agreement might distinguish among types of attacks on space objects (as suggested in Article 3 of the Soviet draft treaty banning deployment in space of "weapons of any kind").

The more fundamental criticisms of the Pressler resolution concern its proposals for a moratorium on ASAT testing and for the elimination of the Soviet ASAT interceptor system. These two provisions are closely linked because a moratorium on ASAT testing alone would perpetuate the current Soviet monopoly of ASAT capabilities. A test moratorium, by itself, would prohibit U.S. advanced development of an ASAT interceptor system while permitting the Soviets to retain their system, which has been successfully tested against satellites in low-altitude orbits (e.g., below 1,000 nm).[75] The main difficulty with the proposal to eliminate existing ASAT interceptors (i.e., the Soviet orbital interceptor system) is that the Soviet interceptor probably could be covertly stored and could rely on support systems from the general Soviet space program for testing and use. For this reason, it probably would be difficult to verify that the Soviet ASAT system had been eliminated.

An additional criticism of these proposed provisions is that they would apply solely to systems "designed exclusively" as ASATs. They would not restrict systems that had ASAT capabilities but that also had other primary missions. For example, the resolution as drafted probably would not affect tests of ground-based BMD lasers against ASATs, because those lasers were not exclusively ASAT systems. Similarly, the Pressler resolution, if agreed by the U.S. and Soviets, would not affect testing or deployment of space-based BMD lasers. If the ABM Treaty were modified or abrogated, an ASAT arms control agreement based on this resolution would not restrict space-based lasers for non-ASAT missions.

Despite the existence of Pressler resolution and the Soviet draft treaty prohibiting space weapons, there is little near-term prospect that an ASAT agreement will be reached. For the reasons noted above (and others), the U.S. probably will regard the Soviet proposal as unacceptable in its present form. Soviet willingness to modify its proposal is unknown, but restrictions on the space shuttle and the inclusion of a right to attack space objects (other than in response to armed attack) could represent Soviet objectives in any agreement limiting ASATs or space weapons generally. U.S. unwillingness to resume bilateral ASAT talks following the Soviet invasion of Afghanistan and the inauguration of President Reagan probably led the Soviets to seek multi-lateral support in the CD. Expressions of interest in ASAT limits by some U.S. allies may also have encouraged the Soviets to try to divide the U.S. from its European allies on this issue. Beyond the specific features of the Soviet draft, the broader subject of ASAT arms control (as indicated by the above criticisms of the Pressler resolution) involves issues that are extremely complex and difficult to resolve. For these reasons, there is little prospect in the foreseeable future that an ASAT arms control agreement would affect development of space-based BMD lasers. (However, because of the length of time projected for such development, prospects for such an agreement would need to be periodically reassessed.)

SUMMARY AND CONCLUSIONS

This chapter has examined the implications of existing arms control agreements and prospective negotiations for development, testing, and deployment of space-based BMD lasers. The chapter focused on the meaning of the ABM Treaty provisions for the development of space-based BMD systems and for fixed, land-based systems "based on other physical principles" than the radar-interceptor systems limited by the Treaty. It then examined prospects for the 1982 ABM Treaty review conference. In later subsections, this chapter examined the implications for space-based BMD lasers of the Outer Space Treaty, other proposed limits on "weapons of mass destruction," and proposed limits on ASATs and on space weapons.

The main conclusion is that the ABM Treaty is the only existing arms control agreement that limits space-based BMD lasers. The ABM Treaty:

- defines "development" on the basis of types of activities that would be detectable by "national technical means" rather than in terms of stages of a weapons acquisition program;

- permits laboratory research and testing of components for space-based BMD systems but prohibits field testing of prototypes of such components against "strategic ballistic missiles or their elements in flight trajectory";

- permits development and "testing in an ABM mode" but not operational deployment of prototype fixed, ground-based BMD laser systems and components;

- permits development and testing of space-based laser components (e.g., pointing and tracking devices) for purposes other than "testing in an ABM mode";

- effectively defines (by U.S. unilateral statement) BMD testing as involving target vehicles whose flight trajectories have the same characteristics as strategic ballistic missile flight trajectories;

- avoids defining "space-based" due to U.S.-Soviet differences on the boundary between outer space and national air space (but, by

U.S. understanding, "space-based" would apply to any orbital object);

- o would not restrict development, testing, and deployment of space-based ASAT lasers, which could also be relevant for BMD laser development (especially because there are no limits on the capabilities of ASAT lasers or on the configurations of target satellites); and

- o in sum, would permit substantial partial testing of components for space-based BMD lasers while prohibiting full-system testing and deployment of such systems.

The *Sentry* BMD development program (as it applies to mobile BMD components) is relevant for space-based BMD lasers because the development, testing and deployment of mobile, land-based BMD systems and components are prohibited by the same provision of the ABM Treaty that also prohibits space-based BMD systems and components. The *Sentry* program underscores the point that substantial *partial* testing of prohibited systems can be done within the terms of the Treaty through different deployment modes. The relevance of the *Sentry* analogy for space-based BMD lasers is not *affected* by the Treaty's agreed statement on BMD systems based on "other physical principles" or because that statement permits development and full-system testing of prototypes. The analogy is mainly limited by the difference in ultimate deployment modes of the systems. *Sentry* testing in a fixed, land-based mode would be very *similar* to testing in a mobile, land-based version. In contrast, BMD testing of ground-based lasers would be significantly different from comparable testing of space-based lasers. However, the *Sentry* analogy indicates that some types of *partial testing* would be permitted. (Similarly, the HOE concept for an exoatmospheric BMD system indicates that testing of sub-orbital BMD laser components might be permitted. However, there is no definitive U.S. position on the compatibility of the HOE concept with the ABM Treaty.)

The *Sentry* example is also important because a decision *in favor* of *Sentry* deployment in a mobile mode would pose a major *test* of the viability of the ABM Treaty. The Reagan Administration probably will not reach a decision on *Sentry* (in the context of permanent MX ICBM deployment) until 1983, and there is no indication that the U.S. will raise the mobile BMD issue, or any other modification of the Treaty, at the ABM Treaty review conference held in the Fall of 1982. If the Administration decides in favor of deployment of mobile Sentry components as part of MX basing, then the U.S. would almost certainly seek amendments to the Treaty to remove the prohibition on land-mobile BMD systems and to increase the limits on BMD deployment for ICBM defense. At that time, and probably depending in part on the Soviet response, the U.S. might also decide whether or not to seek other changes in the Treaty (e.g., to remove the prohibition on space-based BMD systems and components) and also whether or not abrogation of the Treaty would be in the U.S. interest. A U.S. decision concerning the prohibition on space-based BMD systems probably would be heavily dependent on the relative progress of and near-term projections for U.S. and Soviet space-based laser weapons systems.

The Outer Space Treaty is the other existing arms control agreement with possible relevance to space-based BMD lasers. It prohibits deployment in space of "weapons of mass destruction." This term is not defined in the Treaty but is understood by the U.S. (based on a 1948 U.N. resolution) to include nuclear, radiological, chemical, and biological weapons and future weapons whose destructive effects would be comparable to the effects of those weapons, especially nuclear weapons. There is no indication that the U.S. or any other party (notably the Soviets) regards high-energy lasers as included in this prohibition. However, the Soviets did not clearly accept the Western definition of this phrase until a joint U.S.-Soviet statement in 1979 that had the effect of restricting "weapons of mass destruction" to the types of weapons listed above.

During the mid-1970s, the Soviets sought to gain acceptance of a broader concept of the term that would have included lasers (among many other potential weapons) under a comprehensive ban on the development, testing, and deployment of "new types of weapons of mass destruction." However, the Soviet criterion of massive damage to population, environment, or vital facilities would not have applied to space-based BMD lasers. The U.S. consistently opposed the Soviet draft agreement as vague and potentially conflicting with efforts to establish limits for specific types of mass destruction weapons. The Soviet draft has not been withdrawn, but it has received minimal attention since the 1979 joint statement. It is very unlikely that this proposal would be accepted by the U.S. or that it would restrict space-based BMD lasers if it were adopted.

The other type of arms control agreement that could affect space-based BMD lasers would be an agreement prohibiting or limiting anti-satellite (ASAT) systems or other space weapons. Such an agreement would eliminate the possibility of testing and deployment of large space-based lasers that could have BMD capabilities, even though they had not been tested against strategic ballistic missiles. There is no near-term prospect of such an agreement. U.S.-Soviet bilateral talks on ASAT limits were held in 1978-79 but failed to reach agreement and have not been resumed. In August 1981, the Soviets proposed a ban on all space-based weapons, but their draft contains limits on the space shuttle and a right to attack satellites in peacetime that the U.S. probably will oppose. A resolution introduced in May 1981 by Senator Pressler called for an agreement to prohibit ASAT testing and use and to eliminate the Soviet ASAT system. This resolution was successfully held in abeyance by the Administration, and the Pressler proposal would be difficult to implement in part because Soviet compliance probably would be difficult to verify. Although discussions of limits on space-based weapons are being held in the U.N. Committee on Disarmament, those discussions are unlikely to lead to an agreement in the near future.

NOTES

1. U.S. Arms Control and Disarmament Agency, _Arms Control and Disarmament Agreements_, 1982 edition (Washington, DC: USGPO, 1982). p. 140. (Hereinafter cited as _ACDA, 1982_.)

2. U.S. Congress, Joint Committee Print, _Fiscal Year 1982 Arms Control Impact Statements_ (Washington, DC: USGPO, 1981), p. 195n. (Hereinafter cited as _FY1982 ACIS_.)

3. Clarence A. Robinson, Jr., "U.S. to Press MX Deployment During START Talks," _Aviation Week and Space Technology_, June 14, 1982, p. 24.

4. _Military Implications of the Treaty on the Limitations of Anti-Ballistic Missile Systems and the Interim Agreement on Limitation of Strategic Offensive Arms_. Hearings before the Committee on Armed Services, U.S. Senate, 92nd Con., 2nd Sess., p. 39 (June 6, 1982). (Hereinafter cited as _Military Implications_.)

5. Ibid., pp. 40-41.

6. Ibid., p. 275 (June 22, 1972).

7. Ibid., pp. 376-377 (July 18, 1972).

8. Ibid., pp. 439-441, Ryan quoted at p. 441 (July 19, 1972).

9. Ibid., p. 441.

10. Ibid., p. 442.

11. Ibid., p. 443-444, Palmer statement at p. 444.

12. Ibid., p. 171 (June 20, 1972).

13. "U.S. Statement B, tested in an ABM Mode," in _ACDA 1982_, op. cit., 146-147.

14. _ACDA_, 1982, p. 143. Note: The original title ("Agreed Interpretation E") is in _Military Implications_. op. cit., p. 93, and in U.S. Arms Control and Disarmament Agency, _Arms Control and Disarmament Agreements_, 1977 edition, p. 141. (Hereinafter cited as _ACDA_, 1977).

15. Military Implications, op. cit., pp. 40-41.

16. "Letter of Submittal by the Secretary of State to the President on the ABM Treaty and the Interim Agreement," in ibid., pp. 78-97, quoted at p. 81.

17. Foster testimony in ibid., p. 222.

18. Ibid., p. 275.

19. Gerard Smith, Doubletalk: The Story of SALT I. (Garden City, NY: Doubleday, 1980), p. 344.

20. Smith testimony in Military Implications, op. cit., p. 295.

21. Ibid., p. 306.

22. Ibid., p. 371.

23. Ibid., p. 438.

24. Ibid.

25. Ibid., p. 439.

26. Ibid., p. 443.

27. Strategic Arms Limitation Agreements. Hearings before the Committee on Foreign Relations, U.S. Senate, 92nd Con., 2nd Sess., p. 20.

28. FY1982 ACIS, op.cit. pp. 195-196. Note that the comparable statement in the FY1981 ACIS on Ballistic Missile Defense did not specify that the ABM Treaty permits development and testing of BMD systems and components "based on other physical principles" in a fixed, land-based mode. See U.S. Con., Joint Committee Print, Fiscal Year 1981 Arms Control Impact Statements. (Washington, DC: USGPO, 1980), p. 228.

29. ACDA, 1982, op. cit., p. 140. The deployment limits in Article III of the Treaty were further reduced by the 1974 Protocol to the Treaty, which provided that each side could deploy either an ABM system centered on its national capital area or an ABM system for the defense of ICBM silos (rather than both, as provided in the Treaty). See ibid., pp. 162-163.

30. See the discussions of "overlay" BMD system concepts, in U.S. Con., Office of Technical Assessment, MX Missile Basing. (Washington, DC: GPO., 1981), pp. 129-139; and in William A. Davis, Jr., "Current Technical Status of U.S. BMD Programs," in "U.S. Arms Control Objectives and the Implications for Ballistic Missile Defense," (Center for Science and International Affairs, Harvard University, Nov. 1-2, 1979), pp. 29-53 (esp. pp. 35-40). (Hereinafter cited as "U.S. Arms Control Objectives"); See also Colin S. Gray, "A New Debate on Ballistic Missile Defense," Survival, March/April 1981, pp. 60-71.

31. For the text of Agreed Statement E, see ACDA, 1982, p. 141. Note: Agreed Statement E was designated as Agreed Interpretation F in ACDA, 1977, p. 144.

32. Treaty on Outer Space. Hearings before the Committee on Foreign Relations, U.S. Senate, 90th Con., 1st Sess., p. 17 (March 7, 1967). (Hereinafter cited as OST Hearings.)

33. On the air space-outer space boundary issue, cf. Report of the Committee on the Peaceful Uses of Outer Space, General Assembly Official Records, 33rd Sess., (Supplement No. 20 (A/33/20) p. 12. (See also other reports of the Legal Subcommittee of this U.N. committee.)

34. ACDA, 1982, op. cit., pp. 139-40.

35. Ibid, pp. 146-147.

36. Ibid, p. 142.

37. See, for example, "Statement by Secretary of Defense Caspar Weinberger before the Senate Armed Services Committee," Oct. 5, 1981 (Defense Department press release).

38. Washington Post, Aug. 17, 1982.

39. ACDA, 1982, p. 52.

40. U.N. General Assembly Resolution 1884 (XVIII), Oct. 17, 1963, in OST Hearings, p. 132.

41. See, for example, the exchange between Senator Frank Carlson (R., Kansas) and U.N. Ambassador

Arthur Goldberg, in ibid., p. 34 (March 7, 1967).

42. Ibid., p. 23.

43. Ibid., pp. 76-77 (March 13, 1967).

44. Ibid., p. 80 (April 12, 1967).

45. Ibid., p. 100.

46. Ibid., pp. 97-98.

47. Herbert C. Scoville, Jr., "The Arms Control Implications of New Ballistic Missile Defense Technologies," in "U.S. Arms Control Objectives," op. cit., (note 30), p. 110.

48. "Resolution of the Commission for Conventional Armaments" Definition of Armaments," U.S. Doc. S/C.3/30, Aug. 13, 1948, in <u>Documents on Disarmament 1945-1959</u>, Vol. I., 1945-1956. (Washington, DC: USGPO, 1962), p. 176.

49. <u>The SALT II Treaty, Part 6</u>. Hearings before the Committee on Foreign Relations, U.S. Senate, 96th Con., 1st Sess., pp. 208-209. The resolution was subsequently adopted by the U.N. General Assembly. (U.N. Doc 34/79, Dec. 11, 1979).

50. U.S.S.R., "Revised draft agreement on the prohibition of the development and manufacture of new types of weapons of mass destruction and new systems of such weapons" (CCD/511/Rev. 1, Aug. 8, 1977), in <u>Report of the Conference of the Committee on Disarmament</u>, U.N. General Assembly Official Records, 32nd Sess., Supplement No. 27 (A/32/27), Vol II, pp. 1-6. (Hereinafter cited as <u>CCD Report</u>, GAOR A/32/27).

51. U.S.S.R., "On definition of new types of weapons of mass destruction and new systems of such weapons" (CCD/514, Aug. 10, 1976), in <u>Report of the Conference of the Committee on Disarmament</u>, U.N. General Assembly Official Records, 31st Sess., Supplement No. 27 (A/31/27), Vol. II, pp. 276-279, quoted at p. 276-77. (Hereinafter cited as <u>CCD Report</u>, GAOR A/31/27).

52. Ibid., p. 279.

53. Ibid., Vol. I, pp. 41-42.

54. See note 51.

55. U.N. Doc., CCD Report, GAOR, A/31/27, Vol. I, p. 43.

56. CCD Report, GAOR, A/32/27, Vol. I, p. 66.

57. Report of the Conference of the Committee on Disarmament. General Assembly Official Records, Supplement No. 27 (A/33/27). Vol. I, p. 49 (Hereinafter cited as CCD Report, GAOR, A/33/27).

58. CCD Report, GAOR, A/31/27, Vol. I, p. 44.

59. CCD Report, GOAR, A/33/27, Vol. I, p. 50.

60. CCD Report, GOAR, A/32/27, Vol. I, p. 64.

61. Ibid., p. 67.

62. Ibid., p. 65.

63. CCD Report, GAOR, A/31/27, Vol. I, p. 44.

64. CCD Report, GOAR, A/33/27, Vol. I, p. 52.

65. U.N. General Assembly Resolution 34/79 (Dec. 11, 1979), which incorporated the text of the U.S.-Soviet draft resolution introduced in the First Committee (A/C. 1/34/L.7), Nov. 2, 1979. See General Assembly Official Records, Resolutions Adopted on the Reports of the First Committee, 34th Sess. (pp. 80-81).

66. Report of the Committee on Disarmament, General Assembly Official Records, 36th Sess., Supplement No. 27 (A/36/27), pp. 66-75.

67. Ibid., pp. 74-75.

68. United Nations, General Assembly Official Records, 36th Sess., A/36/192 (Aug. 11, 1981).

69. Annex to ibid.

70. The Western resolution is U.S. Doc. A/C.1/36/L.7 (Nov. 10, 1981). The Soviet-sponsored resolution is U.N. Doc. A/C.1/36/L.8 (Nov. 11,1981).

71. U.N. General Assembly, First Committee verbatim record (U.N. Doc. A/C.1/36/PV.39), pp. 47-48 (Nov. 23, 1981).

72. Statement by U.S. Ambassador Louis G. Fields to the Committee on Disarmament, April 3, 1982.

73. Donald L. Hafner, "Arms Control Measures for Anti-Satellite Weapons," International Security, Winter 1980/81, pp. 41-60 (see esp. p. 55). Hafner was a U.S. Arms Control and Disarmament Agency official and served as an analyst with the NSC ASAT Working Group.

74. Congressional Record, U.S. Senate, May 6, 1981.

75. Hafner, op. cit., p. 46.

4
Space-Based Lasers For Ballistic Missile Defense: Soviet Policy Options

Rebecca V. Strode

In seeking to determine the optimum policy toward the development of space-based lasers for ballistic missile defense (BMD) of the United States, it is prudent to consider the range of measures by which the Soviet Union might be expected to respond, and to estimate the likelihood of any particular option being adopted. The implications of this analysis ought then to be measured against the likely development of Soviet policy and research in the field of ballistic missile defense in the absence of a vigorous U.S. BMD program. Research conducted on this issue suggests that, while the Soviets would be highly critical of a decision by the United States to deploy a space-based laser BMD system and would undoubtedly take steps intended to offset or reduce the strategic advantage which such a system might provide, the net effect of such a deployment could be benefical both to the United States' strategic posture and to its foreign policy goals. In any event, given ongoing Soviet efforts in BMD research and deployment, the consequences for the United States of not exploring post-nuclear BMD technologies could well be more onerous than those of even a worst-case combination of probable Soviet reactions to the deployment of a U.S. space-based laser ballistic missile defense.

The methologies used in this study include an analytical survey of the various technological options which would be available to the Soviets in the near-term for countering a space-based laser BMD system, a politico-military analysis of Soviet strategic goals, and a historical study of Soviet behavior in analogous situations in order to identify recurrent patterns in the Soviet response to strategic and technological challenge. Each of these approaches provides unique insight into Soviet priorities and strategic style. Together, they comprehend the problem in all its dimensions -- technological, strategic, and political. To the extent that the conclusions reached by each method of analysis reinforce each other, greater confidence can be placed in the final projection of likely Soviet behavior in the event of a decision in the U.S. either to deploy or to forego laser BMD.

This study is based on a variety of publications. Both Western and Soviet sources have been utilized, in order to gain the most complete picture possible. Among Western sources, this study relies most heavily on unclassified publications and Congressional testimony of the U.S. intelligence community, in the belief that they represent the most informed views on Soviet BMD activities available in the West. But there are considerable gaps even in the data published by the Central Intelligence Agency and Defense Intelligence Agency. Consequently, Western sources are supplemented by an analysis of relevant Soviet strategic and political writings. The value of this literature is limited by the pervasiveness of Soviet secrecy and disingenuousness; nevertheless, a careful reading does provide considerable evidence of Soviet attitudes toward ballistic missile defense.

POSSIBLE SOVIET TECHNOLOGICAL RESPONSES TO U.S. SPACE-BASED LASER BMD

A discussion of possible Soviet reactions to the development or deployment of U.S. space-based laser BMD requires that Soviet calculations be examined within the political context in which they will be made. As signatories to the 1972 Treaty on the Limitation of Anti-Ballistic Missile Systems, the United States and the Soviet Union are obligated by Article V, Paragraph 1 of the Treaty "not to develop, test, or deploy ABM systems or components which are sea-based, air-based, space-based, or mobile land-based."[1] While there are gray areas between research and development which may be exploited within the context of the SALT Treaty, the full-scale development of a system of space-based laser BMD is clearly prohibited. Consequently, the question of Soviet response to a U.S. program to develop or depoly space-based laser BMD presupposes that the United States has withdrawn from, or successfully renegotiated the Treaty in order to accommodate space-based laser BMD. Renegotiation implies some mutuality of interest on the part of the United States and the Soviet Union in the enhancement of strategic defense, and is therefore compatible with the continuation of U.S.-Soviet relations on an even keel. Indeed, the signing of a major new treaty could even signal an improvement in relations between the two powers, much like that which occurred in the immediate aftermath of President Nixon's trip to Moscow in May 1972 to sign the SALT I Accords. A renegotiated ABM Treaty might therefore be obtained without reducing the willingness of the United States and Soviet Union to abide tacitly by the SALT II Treaty's limitations on strategic offensive weapons and without prejudicing U.S. efforts to obtain a follow-on offensive agreement. Withdrawal, on the other hand, implies a unilateral decision by one party to an agreement to undertake an activity which is considered illegitimate, undesirable, or inopportune by the other. Withdrawal from the ABM Treaty is therefore likely to precipitate a general deterioration in bilateral relations between the United States and the Soviet Union, and could alter perceptions in both countries as to the advisability of continued restraint in the deployment of strategic offensive weapons. In this atmosphere of exacerbated relations, the leadership in Moscow may consider a broader range of possible responses than would be the case if U.S. laser BMD development were to take place within the constraints of a renegotiated ABM Treaty. In the discussion of Soviet options which follows, the assumption is made, unless otherwise stated, that the United States has withdrawn from the 1972 Treaty.

Given a U.S. decision to develop or deploy a system of laser BMD, the range of technological responses available to Moscow appears to encompass five discrete options:

a) prevent the United States from mounting an effective laser defense by directly attacking and destroying any space-based laser system which the U.S. deploys;

b) deploy an expanded version of the Soviet Union's current ground-based ABM system, i.e., BMD based on "conventional" nuclear technology;

c) compete with the U.S. in the development of a comparable Soviet space-based laser BMD system;

d) protect Soviet missiles from the effects of lasers, primarily by hardening Soviet boosters;

e) proliferate boosters, warheads, and decoys in an effort to overcome U.S. laser BMD by saturating the system.

Options a), b), and c) constitute different methods of active ballistic missile defense on the part of the Soviets; options d) and e) comprise passive measures of defense.

a) Attacks against U.S. space-based lasers

Of all possible responses, option a) -- the Soviets shooting down any deployed U.S. space-based laser system -- is by far the least likely in peacetime. The risks and the political and material costs of such action would simply be too great. True, the destruction of such a system would not involve U.S. territory or lives, and therefore would not necessarily evoke an immediate U.S. military reprisal. Perhaps the United States would limit itself to a diplomatic protest, in which case the U.S.S.R. would have secured a strategic "payoff" of some consequence and a political victory as well (making the United States look weak). Still, the Soviets could never know for sure what the U.S. reaction would be, particularly if at the time of deployment the United States announced that it considered laser BMD to be an essential component of the nation's strategic deterrent. Certainly by international legal standards, a direct attack on a U.S. strategic asset could constitute a *casus belli*. There is no precedent for an attack on U.S. strategic forces, CONUS-based or otherwise, to guide Moscow in its calculations. How, for example, might the United States respond to an attack on a ballistic missile submarine (SSBN) at sea? Similar uncertainties would surround the laser case. Would the risks be outweighed by the benefits of destroying one replaceable laser system, particularly one which, in its early stage of development, may offer only limited protection to U.S. offensive forces? If, on the other hand, the Soviets initially adopted a "wait and see" attitude, and attacked the orbiting laser weapons only after the system began to reach full deployment or began to look more formidable, their uncertainties would be even greater. By initially acquiescing in the deployment of the first few laser weapons, the Soviets would have helped establish a precedent for the unopposed deployment of space-based BMD. A subsequent attack on such a system would therefore appear all the more unwarranted and bellicose in the United States.

Even if the Soviet leadership were confident that the destruction of U.S. space-based laser assets would not lead to war, they would still have to weigh the benefits of forcefully delaying U.S. attainment of a BMD capability against the costs of the sharp deterioration in U.S.-Soviet relations which would surely ensue. Cooperative endeavors and commercial relations would almost certainly be disrupted in the aftermath of such an attack, exacting penalties across the board for Soviet behavior.

The Soviets might, of course, judge the risk of war arising from the casualty-free destruction of a U.S. laser satellite to be negligible, and they might be willing to accept a considerable deterioration in U.S.-Soviet relations short of war in order to assure the continued viability of their strategic offensive systems. But before the Soviet leadership embarked upon such a policy, it would have to make one more critical calculation. The risk of retaliation would not remain constant over time. At first, it would increase with each incidence of attack; then it would level off and eventually decline as continued U.S. restraint was gradually translated in the minds of both Moscow and Washington into U.S. acceptance of a Soviet <u>right</u> to protect the effectiveness of its offensive forces through direct interference with U.S. strategic deployments (see figure 1). Admission of this principle by the United States would be tantamount to conceding the Soviets hegemony in space. Given the vast resources which the U.S. Government has invested in space programs, as well as the political issues of national status and defense credibility involved, this is a most unlikely course for the United States to follow. The U.S. might allow the Soviets to destroy one laser satellite without reprisal -- perhaps even two or three -- but eventually the pressure to respond would become enormous.

In these circumstances, perhaps the most likely U.S. response would be to attack a Soviet space-based asset -- perhaps a communications satellite or some component of the Soviet space station complex. Such a response would minimize the risk of escalation because it would be bloodless, would not involve Soviet or Warsaw-Pact territory, and would be clearly and directly proportionate to the Soviet activity it is designed to stop. In undertaking to attack U.S. space-based laser assets, the Soviets would have to consider the very real possibility that their actions would escalate into a war of attrition against each other's satellites.

The Soviet Union might have considerable reason to doubt its ability to compete with the United States in an expensive "space war" of this type. By the late 1980s to early 1990s, when laser BMD systems might first begin to be deployed, both the Soviet Union and the United States will have anti-satellite weapons, but there is evidence that the U.S. systems may exhibit superior performance. The current Soviet system consists of a large booster carrying a rocket-propelled homing vehicle which is placed in orbit before attacking its target. While the system is currently limited to low earth orbit, development is underway of a larger booster to provide the greater

FIGURE 1. LEVELS OF RISK ASSOCIATED WITH SOVIET ATTACKS ON U.S. LASER ASSETS OVER TIME

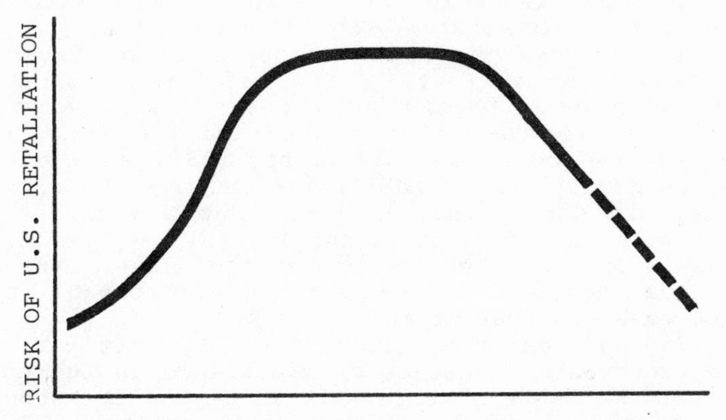

velocity required to reach geosynchronous orbit. The U.S. ASAT system currently under development will consist of a Short-Range Attack Missile (SRAM) and a miniature homing device launched from an F-15. This system, with a planned initial operational capability (IOC) of 1985, will boost not into orbit, but into a direct-ascent intercept of its target satellite. In principle, this system should have a lower cost, shorter response time, and higher rate of fire than its Soviet counterpart.[2] Even if there is a two-year slippage in the IOC of the U.S. ASAT system, it will still be available by the time the first laser BMD test bed could be established (1987 at the earliest). Soviet ASAT capabilities can also be expected to improve during this period. But a straight-line projection of current technological sophistication suggests that by the time space-based laser becomes a reality, the U.S. will possess an ASAT capability at least equal to that of the U.S.S.R.

In addition, the United States by this time will have accumulated a decade of experience in the operation of the space shuttle. It is beyond the scope of this study to examine the range of military uses to which the shuttle might be put, but the Soviet Union clearly perceives it as a military threat. A February 1982 article in the journal of the Soviet Academy of Sciences' Institute for World Economics and International Relations asserts that "the shuttle program is being subordinated more and more to the interest of the military,"[3] and notes in particular that "the possibility of using the shuttle fleet to help develop anti-air (PVO) and anti-missile defenses (PRO), equipped with laser and beam weaponry is being studied."[4] Even if the Soviet Union deploys its own version of the space shuttle in the late 1980s or early 1990s, it is unlikely to attain immediately the levels of reliability and performance which will accrue to the more mature U.S. system. The Soviets are therefore likely to suspect, with considerable justification, that they would be the principal losers in a prolonged anti-satellite exchange. This prospect might well provide the decisive argument in Soviet calculations against directly attacking U.S. space-based laser assets in peacetime.

Quite apart from the immediate risks involved in attacking U.S. space-based systems, there are barriers to such a policy which derive from Soviet historical experiences and strategic doctrine. None of the wars in which the Soviet Union has engaged over the course of its history has been caused by technological developments in another state. Technological challenge has spurred the development of more sophisticated weaponry, larger procurements, and changes in tactics, but it has not served as a rationale for aggression. On the contrary, it has occasionally provided the leverage required to draw the Soviets into serious arms control negotiations. The most striking example of this phenomenon is the Soviet participation in SALT I negotiations. The implications which Soviet behavior in SALT I may hold for Moscow's response to U.S. laser programs are addressed in Section III of this study.[5] Another example might include the current Intermediate-range Nuclear Forces (INF) negotiations currently underway, through which the U.S.S.R. is seeking to minimize the impact of the cruise missile and the Pershing II ballistic missile on the European theater balance.

For the Soviet leadership, war is an instrument of politics, not a function of technology. General V.D. Sokolovskii, the premier Soviet strategist of the post-war era, has criticized the notion that a technological development could overrule the dictates of political prudence. Such thinking, he claimed, is

> the consequence of a metaphysical and unscientific approach to such a phenomenon as war, and a result of idealization of the new weapons. It is well known that the essence of war as a continuation of politics does not change with changing technology and armament.[6]

More recently, Colonel V.M. Bondarenko, a prominent Soviet commentator on the role of technological development in the armed forces, underscored the primacy of strategy over technology, and politics over strategy. "Military science," he wrote,

> is at one and the same time both a 'filter' which sorts the achievements of science and technology according to their relevance for military affairs, and a 'magnet' which attracts scientific research toward the solution of problems concerning the country's defense capability. Of course, the effectiveness of this system depends not only on military science. Here an enormous role is played also by policy, which creates the social requisites for strengthening the ties between the military and science, and which illumines these ties with moral goals and motives.[7]

While there is nothing in Soviet strategy or history to suggest that the U.S.S.R. would risk a direct attack on a high-priority U.S. strategic asset in the absence of extreme political tension and for no reason other than to simplify its own strategic calculations, such an attack should be expected in the event of a decision by the Kremlin actually to embark upon nuclear war. The offensive emphasis of Soviet strategy places a premium on limiting damage to the homeland by achieving a debilitating first-strike against enemy retaliatory forces. To attain this end, tactical surprise is essential. A 1976 article on "surprise" in the authoritative _Soviet Military Encyclopedia_ elaborates this concept by noting that "surprise is one of the most important principles of military art," and that its chief aim is "to permit delivery of a strike when the enemy is least prepared to repulse it and thereby to paralyze his will for organized resistance."[8] "By making it difficult for the enemy to take retaliatory measures," the article continues, "surprise makes it possible to achieve maximum results with the least expenditures of forces, weapons, effort, and time."[9]

An effective U.S. ballistic missile defense capability would severely constrain Moscow's ability to reduce the effectiveness or destructiveness of a U.S. retaliatory strike, and would thereby deny the Soviets a fundamental tenet of their victory conditions (damage-limitation). Consequently, space-based laser weapons would constitute a high-priority, time-urgent target set in any Soviet first-strike contingency plan for nuclear war.[10] Prudence would therefore suggest that a space-based laser system be provided with some measure of protection. Since a system of space-based laser BMD is a defensive system and is not a weapon of mass destruction, its deployment would not constitute an immediate or severe threat to the survival of the Kremlin leadership, and therefore a "bolt-from-the-blue" attack against such a system is highly improbable. Even if the combination of space-based BMD and counterforce-capable offensive weapons were to provide the United States with the theoretical capability to execute a preemptive strike against the Soviet Union while relying on ballistic missile defense to limit the damage incurred for the U.S.S.R.'s retaliatory strike, the risk of technical failure in such a plan is so great, and the consequences of failure are so severe, that the Soviet Union could confidently discount the possibility of an unprovoked U.S. attack. And if the Soviets do not fear an unprovoked attack, they would have no incentive to provoke one by destroying U.S. strategic assets. Thus, one is again led to the conclusion that in the absence of a decision to initiate nuclear war on the basis of political concerns unrelated to the presence or absence of U.S.

BMD, the Soviet Union is most unlikely to attempt the destruction of a U.S. space-based laser BMD system.

A far more credible Soviet response would be to prepare means in peacetime by which U.S. space-based laser ballistic missile defense could be rendered ineffective in time of war. To do so, the Soviets might choose to rely on any of the options b) through e) outlined above, but would be most likely to choose some combination of several or all of these countermeasures. First, a variety of defense would complicate U.S. efforts to devise counter-countermeasures. Second, there may be substantial synergisms to be gained from a combination of both active and passive means of countering threats to the ICBM force.[11] Third, the Soviets' current strategy of damage- limitation incorporates active and passive elements. The active pursuit of damage-limitation is evinced by the capability of the Soviet ICBM force to destroy in a first-strike some 90 percent of U.S. fixed, land-based ICBMs, a substantial proportion of U.S. strategic bombers, and all of the ballistic missile submarines (SSBNs) in port. The impact of a retaliatory strike by remaining U.S. bombers, submarines, and ICBMs might then be blunted by civil defense efforts. However, effective this strategy may or may not be, it indicates Moscow's appreciation of both forms of defense, and suggests that a similarly combined approach might be adopted in response to U.S. laser BMD.

b) <u>Deployment of an expanded ground-based ABM</u>

In the face of a BMD challenge to the effectiveness of the U.S.S.R.'s strategic offensive forces, the Soviet Union could respond with a vigorous ballistic missile defense program of its own. Such a program would not by itself constitute a restoration of the strategic *status quo ante*, because it would do nothing to enhance the penetrability of Soviet missiles. At present, the U.S.S.R.'s powerful ICBM and SLBM forces appear to have two wartime missions -- one offensive, the other defensive. Offensively, Soviet missile forces would be used to obtain "the complete rout of the enemy's armed forces, the destruction of his economic potential, the disorganization of the system of political and military control, and the capture of enemy territory."[12] Ultimately, Soviet counterforce and counterpolitical targeting appears to be directed toward the elimination of the enemy's ability to resist Soviet post-war demands by destroying his military capacity and political cohesion. As

Marshal I. Kh. Bagramian has explained, "Massive nuclear strikes by strategic weapons allow the attainment of the political aims [of the war] in a short period of time."[13] By rendering Soviet strategic forces vulnerable in flight, U.S. space-based laser BMD could significantly degrade the U.S.S.R.'s capability to achieve the offensive goals of its strategy, and the development of a comparable Soviet BMD capability would do nothing to restore this lost potential.

There is, however, an important defensive role for Soviet missile forces. By destroying U.S. forces before launch, the U.S.S.R. can hope to limit damage to its own homeland, thereby enhancing its potential for war survival and post-war recovery. Lieutenant General I. Zavialov noted this aspect of Soviet strategy in 1970. "Nuclear weapons," he said, "have increasingly confirmed the role of attack as the decisive form of military action and have given rise to the necessity of resolving even defensive tasks by active offensive actions."[14] An effective U.S. ballistic missile defense would undermine Soviet damage- limitation efforts by protecting U.S. retaliatory forces. But a comparable Soviet BMD capability could compensate for the loss of a damage-limiting role for Soviet missiles by protecting Soviet assets even after U.S. missiles are launched.

This in itself might well provide sufficient argument within the Soviet General Staff and Politburo for matching a U.S. deployment of BMD with an enhanced BMD system of their own. The peacetime political benefit of showing oneself capable of meeting any U.S. challenge would further enhance the system's attractiveness. The Soviets have expressed the need to prove that they are the technological equals (or superiors) of the United States in weapons development. Soviet writings point with pride to their achievements in the field of nuclear warheads and delivery vehicles, claiming to be the first nation to develop a deliverable hydrogen bomb and pointing to their early achievements in the development of ballistic missiles.[15] With regard to the possible deployment of the enhanced radiation warhead in Europe, Brezhnev reiterated his determination to develop an adequate response:

> The Soviet Union is resolutely opposed to the creation of the neutron bomb...But if this bomb is created in the West--created against us, something that no one even tries to hide--the West should clearly realize that the U.S.S.R. will not remain a passive observer. We shall be confronted with the

> need to reply to this challenge in order to ensure the security of the Soviet people and of their allies and friends. In the final analysis, all this will raise the arms race to a still more dangerous level.
>
> We do not want this, and therefore we suggest that an agreement be reached on mutual renunciation of the production of the neutron bomb...[16]

The clear implication is that a U.S. ER warhead would be matched by a Soviet one. The same logic might apply to space-based laser development. There would be a strong desire among Kremlin officials not to be left behind, or not to appear to be left behind, in a weapon technology which promises substantial long-term strategic benefits.

While the Soviets would probably seek to deploy their own laser BMD system eventually, their initial response to a U.S. laser BMD deployment might emphasize "conventional" nuclear-armed, ground-launched ABM interceptor missiles. For example, the Soviets might rapidly deploy an expanded and modernized version of the ABM system already in place around Moscow. Given the continuing research on ABM radars and interceptor missiles which the U.S.S.R. has conducted over the past decade, the Soviets might hope that the rapid deployment of such an upgraded system, based largely on previously assimilated and tested technology, would provide the U.S.S.R. with an early, short-term "lead" in deployed BMD capability while buying time for the development of a more capable follow-on system of space-based laser defense. There is evidence that the U.S.S.R. is already laying the ground work for such an option. Dr. Jack Vorona, the Defense Intelligence Agency's Deputy Director for Scientific and Technical Intelligence, testified in November 1979 that to counter sophisticated threats, the Moscow ABM system, which appears to have a range of some 200 nautical miles and is probably armed with a 1-2 megaton warhead,

> will require significant modifications and/or replacement, including discrimination radars capable of handling large numbers of targets and high performance interceptor missiles. There is every indication that the Soviets are pursuing R&D supportive of these requirements.[17]

In 1980, the U.S.S.R. deactivated half of its 64 ABM-1B (Galosh) interceptors, leaving the other 32 still in place around Moscow. Many Western defense analysts believe this deactivation was in preparation for upgrading the entire system.[18] A new ABM-X-3 radar has been under development for a decade.[19] The U.S. intelligence community now has evidence that the U.S.S.R. has developed and is producing components for a BMD system comprised of transportable phased-array radars and hypersonic interceptor missiles. Some elements of the new technology developed for this system have already been applied to the existing Galosh ABM system around Moscow.[20] The Central Intelligence Agency has observed no evidence that the Soviet Union is currently planning to abrogate the ABM Treaty or to deploy a nationwide ABM system. However, CIA military economics analyst Donald Burton contends that the Soviets certainly have the economic capability to deploy a nationwide system by the mid-1980s.[21]

The new ABM launcher and missile which are apparently being prepared to replace the thirty-two deactivated interceptors around Moscow are reportedly in a production line which normally turns out much larger quantities of equipment than would be required to modernize the Moscow site alone. This suggests that the Soviets have placed themselves in a position to move quickly to a larger system, at a time of their choosing.[22] Thus, as Colonel William J. Barlow of the U.S. Air War College explains, Soviet ABM modernization "provides the basis for a potential 'breakout' threat."[23] Whether the Soviets decide to utilize that potential will depend on their perceived level of need or opportunity, which in turn is partly a function of U.S. BMD activity. Deployment by the United States of space-based laser BMD could provide the impetus for a decision in Moscow.

Could such a ground-based, nuclear-armed ABM system be effective? Clearly the current system, even if deployed to the maximum permissible level under the SALT Treaty of 100 launchers, could be easily overwhelmed by U.S. strategic forces, even if the Soviets are granted the first-strike. But a nationwide, multi-site deployment of an upgraded ABM system could prove marginally effective against a first-strike by the United States through the 1980s, and might provide substantial damage-limitation against the much reduced retaliatory forces which would be available to the United States following a Soviet first-strike. The employment of the Soviet Union's 308 SS-18s alone is probably sufficient to destroy most of the United States' currently deployed ICBM forces, and an extensive ABM deployment could be effective in reducing the penetrability of the 10-50 Minuteman ICBMs which might survive the initial onslaught.[24] Thus, until the MX ICBM is deployed in a survivable basing mode -- that is, until the late 1980s or early 1990s -- a ground-based Soviet ABM system, in combination with Soviet offensive forces, could probably provide significant protection from U.S. ICBM attack.

Of course, the U.S.S.R. would also have to contend with U.S. SLBM and bomber forces, but here, too, a large, nuclear-armed ABM system could make some contribution to damage-limitation. Due to their shorter ranges, Polaris and Poseidon SLBMs generally have slower reentry speeds than do ICBMs, and may therefore be more vulnerable to ABM defense. It is interesting to note in this regard that Kapustin Yar, the launch center for the Soviet MRBMs which serve as targets for the Sary-Shagan ABM test range, is located some 1200 miles from Sary-Shagan. Thus the flight distance and trajectory of these target MRBMs approximate those of the Polaris SLBM.[25] As the Trident I, and later Trident II, submarines come on-line throughout the 1980s, the anti-SLBM capability of a Soviet ABM system would presumably diminish. In addition, twelve Poseidon submarines are now programmed to be refitted with the extended-range Trident I (C-4) missile.[26] An enlarged "conventional-style" ABM system could probably only provide a SLBM defense for a short period of time -- say, until the late 1980s. But, as has been suggested, the Soviet Union would probably consider such a deployment to be only a short-run BMD solution anyway.

In addition to the Moscow ABM system, the U.S.S.R. maintains a large anti-aircraft defense network of interceptor aircraft and some 10,000 surface-to-air missiles (SAMs). There is evidence that the Soviet Union has considered utilizing certain of their SAMs in an ABM role. In 1973 and 1974, for example, the United States observed ballistic missile flights made in conjunction with SA-5 radar tracking.[27] The SA-5 can intercept targets at up to 100,000 feet at a slant range of 150 miles.[28] While the actual ABM capability of this system is probably marginal at best, future "SAM upgrades" could be somewhat more effective. The U.S. intelligence estimate of the main mission of the Soviet Union's new, hypersonic SA-11 is classified. This two-stage, solid-fuel missile may only be a tactical anti-aircraft missile, perhaps a follow-on to the SA-6.[29] But the suggestion has been made that it could be intended as an anti-cruise missile or an ABM missile.[30] This may be unfounded speculation; still, in considering possible short-run Soviet responses to U.S. BMD deployment, the use of improved surface-to-air missile systems in an ABM mode must be considered. By itself, such deployment would probably provide only minimal protection from incoming U.S. warheads. But used in conjunction with a large, nationwide ABM system, upgraded SAMs might add a measurable degree of defense capability.

In this discussion of the utility of ground-based ABM interceptors for ballistic missile defense in the 1980s, the assumption has been made that such a defense would be effective only if a large proportion of U.S. forces were eliminated prior to launch by a preemptive Soviet first-strike. The U.S.S.R. either now possesses or soon will possess the technical capability to execute such a strike.[31] But the assumption has also been made that this ABM deployment would come in response to a U.S. decision to deploy space-based laser BMD. The question then arises whether even an improved and enlarged *Galosh*-style ABM system could be effective against a U.S. retaliatory strike by forces which have been protected by space-based laser BMD. If space-based laser BMD proves highly effective in protecting U.S. strategic missiles, it seems likely that a Soviet ground-based ABM system based largely on currently assimilated technology could be overwhelmed.

However, it will take time for the alternative, space-based laser BMD, to reach full deployment. Assuming program initiation in 1983, the first U.S. test bed might be ready by approximately 1987, and full deployment would not be expected before the early to mid-1990s. The Soviet Union could therefore expect to have some five to eight years subsequent to the initiation of a U.S. laser BMD development program before Soviet offensive missiles would face a fully operational BMD threat. As U.S. laser BMD satellites were deployed in the late 1980s and early 1990s, Soviet preemptive capabilities would attenuate, but this would be a gradual process, not a sudden degradation. In the meantime, Soviet forces would still be capable of inflicting substantial losses on U.S. ICBMs, bomber bases, and inport submarines, thereby reducing the U.S. retaliatory threat to more manageable levels.

With the required technology already in hand, the U.S.S.R. might judge large numbers of ground-based ABM interceptors deployed at multiple sites to be a cost-effective means of dealing with this interim BMD challenge. First, it would offer moderate, but not inconsequential, damage-limitation. Secondly, because it could probably be deployed fairly rapidly -- i.e., before the U.S. fully deployed a laser BMD -- it would offer the political advantage of appearing to demonstrate Soviet superiority in BMD capability. Thirdly, such a deployment would mesh well with the Soviet Union's traditional emphasis on incremental improvements in weapons design and procurement.[32] And finally, a nationwide ground-based ABM system would provide a hedge against possible failure in the development of unproven technologies associated with space-based laser BMD.

c) <u>Deployment of a Soviet Space-Based Laser BMD</u>

While ground-based ABM may be viewed as an attractive short-term response to the development of space-based laser BMD by the United States, the greater long-run strategic promise of laser BMD is likely to lead the Soviet Union into some form of unconventional (e.g., space-based laser) deployment of its own. As U.S. laser satellites came on-line, the political and strategic pressures on the Kremlin not to concede an asymmetrical damage-limitation capability to the United States would mount; and these pressures, combined with the considerable interest the Soviet Union has already demonstrated in unconventional BMD technologies, again suggest that Soviet deployment of ground-based, nuclear-armed ABM interceptors will be accompanied by accelerated efforts to develop more sophisticated BMD capabilities for the future.

The Soviet Union is currently exploring three unconventional technologies for BMD application: laser systems, particle beam weapons, and millimeter waves or electromagnetics. In the latter area, the Soviets are actively developing the very high peak power microwave generators relevant to such application.[33] The Defense Intelligence Agency (DIA) believes that particle beam weaponry technological capabilities are roughly equivalent in the United States and the Soviet Union.[34] These avenues may continue to be explored, but laser technology may appear to hold the greatest BMD promise.

To the extent that the U.S.S.R. judged space-based laser to be technically feasible and strategically attractive, the Soviet research and development effort could be expected to be large and relatively fast-paced. Already, the Soviet Union's high energy laser program is about four times the size of its U.S. counterpart.[35] According to former Undersecretary of Defense Research and Engineering William Perry, the Soviet laser program differs from the U.S.S.R.'s typically conservative philosophy of gradual technological advance through progressive modifications programs. Rather, Soviet laser development has been characterized by a great amount of innovation, instigated by "high level policy intervention".[36] In this respect, the laser program resembles earlier Soviet high-priority programs, such as nuclear weaponry and ICBM development.[37] While it is not the purpose of this chapter to make a precise estimate of the time frame which would be required by the U.S.S.R. to develop such a system, it should be recalled that previous "crash programs" on such high-priority military systems as the atomic and hydrogen bombs and MIRVed ICBMs have produced results within a relatively few number of years, and considerably more quickly than had been predicted by many Western intelligence analysts.[38] According to a 1982 report in Aviation Week and Space Technology, "the long-term outlook for Soviet space program advancement is for continued aggressive research and development, with the Soviets bringing military space systems on line after less development time than comparable U.S. systems".[39] Perhaps the most succinct Western estimate of Soviet laser BMD potential is that of DIA's scientific intelligence analyst Jack Vorona in Congressional testimony. Dr. Vorona declined to specify in open (unclassified) session whether the U.S.S.R. is now building a prototype space-based laser BMD system. He did, however, state that

> the U.S.S.R. appears to be roughly comparable to the United States in the capability to develop high energy laser systems. They have been working on the basic laser technologies as long as the United States, and apparently have the expertise, manpower, and resources to develop any type of weapon laser that the United States could.[40]

Vorona's assessment, however, is not universally accepted. According to a 1982 study by the General Accounting Office, the United States leads the U.S.S.R. in many technologies related to space-based laser by five to ten years. These include optical systems, miniaturization, computers, and lightweight spacecraft. The Soviet lag in microelectronics and computer technology is estimated to be from two to seven years.[41]

Whether the Soviet technological lag is in fact a relatively short two years or a relatively long seven is of course an important factor in Soviet calculations of their own competitive position in the development of laser BMD. If the Soviet leadership were to determine that the capabilities of Soviet science and industry to produce such a system were grossly inferior to those of the United States, the U.S.S.R. might well attempt to solve its strategic dilemma through diplomatic means -- hoping even after the collapse of the original ABM agreement to obtain new Treaty restrictions which would prevent the United States from deploying an effective space-based BMD system. Even if the Soviets are prepared to deploy their own system, they may still seek some sort of renewed BMD negotiations in order to strengthen their political and strategic position vis-a'-vis the United States. This prospect is addressed in greater detail later in this chapter.

d) Passive Protection of Missiles from Laser Effects

A variety of passive measures for defending Soviet missiles from laser effects might be adopted. ICBM and SLBM boosters might be hardened in order to improve their survivability under laser attack. Richard Garwin estimates that a booster hardness of two kilojoules per square centimeter would impose a cost on the ballistic missile defender of approximately $100 million per booster destroyed. By reducing by one or two the number of warheads carried on any one missile (i.e., "off-loading" one or two MIRVs per missile), ICBM and SLBM boosters' surfaces could be hardened against laser energy by a factor of perhaps ten or more, to between 10 and 20 kj/cm^2.[42] The large payload of Soviet fourth-generation ICBMs will probably be maintained at roughly the same level in the fifth-generation follow-on systems now under development.[43] The potentially greater throw-weight of these missiles (relative to the Minuteman and the MX), and the greater reentry vehicle (RV) loading capacity which it provides, might convince the Soviets that a small reduction in RVs would be an acceptable price to pay for improved boost-phase survivability.

Other defensive measures could also be taken. For example, Garwin has suggested that boosters might be protected from laser illumination by "hiding" behind aluminum foil screens perhaps 100 meters square. Missiles could also be equipped with electronic countermeasures (ECM) for use against the laser system's command and control.[44] Kosta Tsipsis has suggested three other defensive measures which he contends would provide highly cost-effective means for protecting missiles from the effects of laser light. The surface of the missile could be made highly reflective so that little light is absorbed. A missile could be provided with an ablative coating, which would burn off and carry with it the energy of the laser radiation. A layer of fluid continuously secreted from the nose of the missile could produce a similar effect.[45] The Soviets might experiment with a number of these options.

e) Proliferation of Boosters, Warheads, and Decoys

The Soviet Union could attempt to overcome U.S. space-based laser BMD not by protecting missiles against laser weapon illumination, but rather by overwhelming the BMD system with large numbers of targets. By greatly proliferating the number of boosters, an attacker could theoretically saturate the target acquisition and aiming capacities of the opposing laser satellites. The Soviets might therefore hope to retain a significant hard-target "leak through" capability even in the face of a fully deployed U.S. laser BMD system, particularly if they simultaneously capitalize on their advantage of greater throwweight by further fractionating their warheads so as to maximize the lethality of each surviving missile. Saturation of a laser weapon system could theoretically be obtained even without proliferating the number of deployed boosters if a large proportion of current missile forces were launched simultaneously. This tactic, however, exacts important strategic penalties. First, by reducing the attacker's inventory of missiles held in reserve, the attacker may lose post-strike bargaining leverage. Secondly, the effectiveness of his targeting may be reduced by limiting his ability to employ a "shoot-look-shoot" tactic (i.e., by reducing his ability fully to exploit his damage-assessment capabilities). It therefore seems probable that if saturation is attempted, proliferation of boosters will be the chosen method.

At the present time, there are no international legal prohibitions against proliferation of boosters by the Soviet Union or by the United States, as the SALT I Interim Offensive Agreement has expired and a follow-on accord has yet to be ratified. Thus far, both the United States and the Soviet Union appear tacitly to have agreed to comply with the terms of the SALT II accord as though it had legal force, but should the United States withdraw or, through renegotiation, substantially relax the terms of the ABM Treaty, the Soviets might find voluntary compliance with a moot SALT II Treaty no longer in their interests. Because withdrawal from the ABM Treaty implies a more hostile political environment than renegotiation, Soviet resort to missile proliferation appears more likely to follow a unilateral decision by the Untied States to withdraw from the ABM Treaty than to accompany a joint U.S.-Soviet agreement to re-negotiate the Treaty. Nevertheless, it could occur in any event.

It is possible that before a decision to respond to U.S. laser BMD becomes imperative for the Politburo, a new strategic offensive arms limitation agreement will have been negotiated and ratified. A hypothetical "SALT III" would probably restrict Soviet flexibility in proliferating boosters, particularly in light of the Reagan Administration's professed commitment to numerical reductions in strategic systems.[46] At this point, though, there is little prospect for an early agreement. Consequently, a Soviet decision to proliferate boosters and warheads or not is likely to be made in the absence of any formal treaty restrictions on Soviet options.

It should be noted that while options b) through d) (ground-based ABM deployment, laser BMD deployment, and passive or ECM resistance) are mutually compatible, options d) and e) are partially exclusive. One cannot simultaneously off-load MIRVs and increase warhead fractionation. Of course, warhead fractionation need not accompany booster proliferation. One could off-load MIRVs as envisioned in option d) in order to harden boosters and at the same time proliferate the number of boosters, but the lethality of each surviving missile would be reduced. Alternatively, the Soviets could harden certain missile boosters (and correspondingly reduce the number of RVs such missiles carried) while leaving other boosters unhardened (and fractionating those missiles' warheads in order to obtain the maximum number of hard-target capable RVs per surviving missile). Accuracy improvements which should be forthcoming over the next decade will make greater fractionation possible without sacrificing hard-target kill capability.

EVALUATING THE OPTIONS

Option a), direct attack against U.S. space-based laser BMD assets, entails such grave risks and potentially enormous costs as to appear an unlikely choice in peacetime, particularly in light of the numerous other, less risky and less expensive, options available to Soviet planners. Options b) through e) all constitute credible responses to U.S. laser BMD, though they undoubtedly vary in their costs and effectiveness. It has been suggested that the Soviets might best enhance the penetrability of their missile forces through some combination of these last four options. But given the large budgetary outlays which certain responses, particularly c) and e), would entail, the question arises whether the Soviet Union can afford to adopt all these measures, or if so, to what degree. No estimate of the ruble or dollar costs which Soviet BMD deployment programs might entail is available, but clearly they would be high. Combining BMD options b) and c) with the pursuit of much enhanced and enlarged offensive force procurements as envisaged in option 3) would prove enormously expensive.

There is no question but that the Soviet Union is willing to bear a greater defense burden than the United States, both in absolute terms and as a share of GNP. In 1980, for example, the total cost of Soviet defense activities was approximately 50 percent higher than the U.S. total in dollar terms, or approximately 30 percent higher than U.S. costs in ruble terms.[47] Soviet defense activities accounted for some 11 to 12 percent of Soviet GNP throughout the 1970s, while U.S. defense activities accounted for approximately 8 percent of U.S. GNP in 1970 and 5 percent in 1979.[48]

Within these aggregate levels, expenditures for military investment, RDT&E (research, development, testing and evaluation), and strategic offense and defense are particularly germane to this discussion. The estimated dollar cost of Soviet military investment exceeded comparable U.S. outlays by 55 percent over the 1970s, and by 80 percent in 1979.[49] If military RDT&E, procurement, and construction are combined, one finds that the Soviet Union started the 1970s with outlays in these categories approximately equal to those of the United States. But the Soviet Union increased its spending at a rate of about 4 percent per year throughout the decade, while the United States decreased its real outlays each year until 1975. As a result, the U.S.S.R. invested over the decade about $240 billion (in FY1981 dollars) more than the United States. This differential exceeds the estimated acquisition cost (in 1981 dollars) of 1000 F-16s, 1000 F-18s, 10,000 XM-1 tanks, 20 CG-47 guided missiles cruisers, 50 nuclear-powered attack submarines, 20 TRIDENT I submarines with missiles, and the entire MX program, while still leaving an additional $70 million in R&D.[50] As for strategic offensive outlays, Figure 2 graphically depicts the far greater magnitude of Soviet expenditures. Finally, Soviet outlays for strategic defense remained high throughout the 1970s, while comparable U.S. activities declined. As a result, the dollar cost of Soviet strategic defense activities increased from 5 times

FIGURE 2. U.S. AND SOVIET STRATEGIC OFFENSIVE EXPENDITURES[51]

US AND SOVIET FORCES FOR INTERCONTINENTAL ATTACK

A comparison of US outlays with estimated dollar cost of Soviet activities if duplicated in the United States

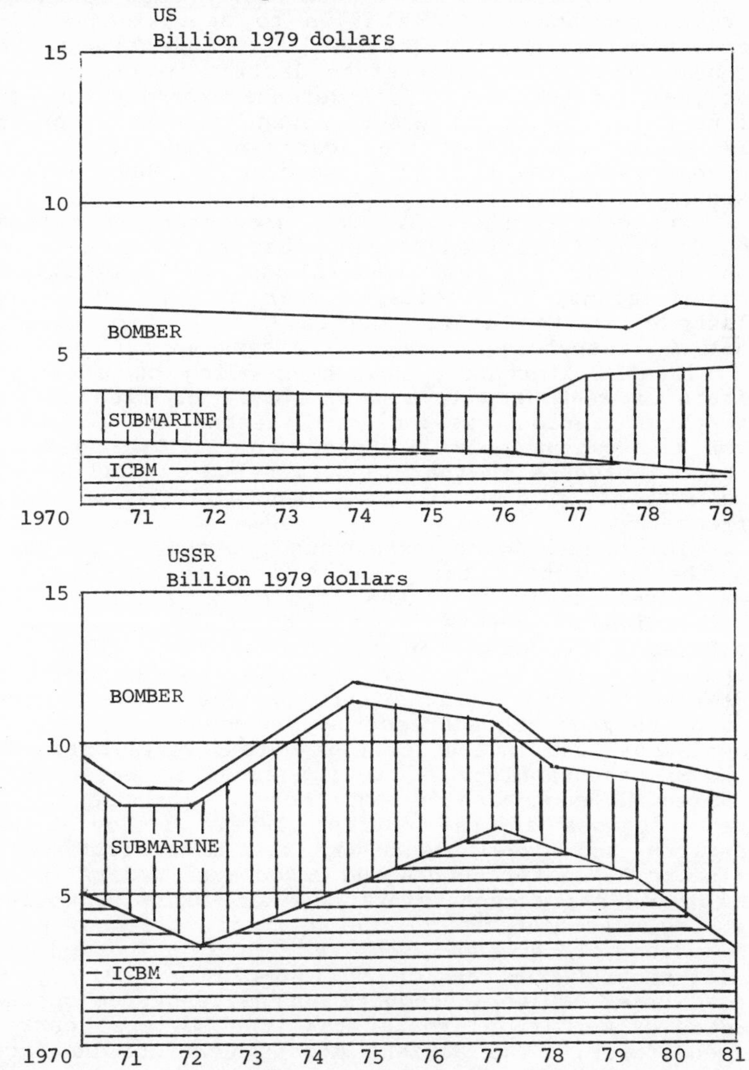

NOTE: The intercontinental attack mission is defined according to the US Defense Planning and Programming Categories of November 1979, with minor adjustments made to attain comparability. Cost for pensions, nuclear materials for warheads, and RDT&E of both sides are excluded. The peripheral attack forces of the USSR are also excluded.

U.S. outlays in 1970 to 25 times U.S. outlays in 1979.[52]

While Soviet defense spending has grown steadily over the past decade at a rate of 4-5 percent per year, Soviet economic growth has slowed from roughly 4 percent per annum in the 1970s to approximately 2-3 percent per annum in 1980-1982. In 1983-1985, Soviet economic growth is expected to decline to 1-2 percent per year. If the U.S.S.R.'s defense expenditures continue to rise at roughly 4-5 percent per annum, as the CIA anticipates, then the share of GNP devoted to defense will rise from 11-12 percent in 1980 to 13-15 percent in 1985.[53]

Yet despite the U.S.S.R.'s slow economic growth, the Central Intelligence Agency has found no evidence that economic problems have slowed the momentum of Soviet defense activities, either at the production plant or in the design bureau.[54] DIA anticipates that while spending on defensive systems will undergo a "sizeable increase", Soviet spending on strategic offensive weapons will remain steady or rise as new (fifth-generation) weapons are tested and deployed over the coming years. Consequently, DIA does not expect increases in the U.S.S.R.'s budget for defensive systems to come at the expense of the strategic offense.[55]

This intelligence consensus appears to be based on the assumption that Soviet strategic priorities remain relatively insulated from major shifts in the correlation of forces brought about by U.S. weapon systems. Prospects for the budget-share of Soviet strategic-defensive programs appear to be based on relatively straight-line projections from recent years; they do not address the consequences of a discontinuous leap in the level of defensive effort. But if the penetrability of Soviet missiles were to be challenged, such a shift might occur. Although moderate increases in the current level of Soviet BMD research and development and the modernization of existing ABM interceptors and associated radars within the 1972 Treaty constraints may not force reductions in budget allocations to the strategic offense, major expansion of ground-based ABM into a nationwide, multi-site system and/or full deployment of space-based laser BMD would require defensive-system outlays orders of magnitude greater than the Soviets' current expenditures. The Soviets might therefore be faced with trade-offs between offensive and defensive systems, leaving Moscow with tough decisions and perhaps sub-optimal solutions.

Assuming these relatively stringent budget constraints, it is doubtful the U.S.S.R. could simultaneously pursue all four of the "desirable" options outlined above, i.e., options b) through e). In particular, there would seem to be a trade-off between the relatively high-cost, "offensive" approach outlined in option e) (booster proliferation) and the "defensive" approaches comprising options b) and c) (ground-based ABM and space-based laser BMD). Exclusive choices need not be made, but it seems likely that heavy reliance on booster proliferation would imply considerably less emphasis on the competitive BMD approaches, and vice versa. Some of the passive protection measures subsumed under option d) would probably be pursued in any case, as they appear to be relatively inexpensive.

Booster proliferation might offer the simplest response to U.S. laser BMD, but because sole or predominant reliance on this approach would concede to the United States the strategic and political advantages of a potentially decisive technological lead in BMD, it is not likely to be given highest priority. The Soviet Union would probably opt instead for a combination of options b) through d) -- i.e., ground-based ABM for the short-term, space-based BMD for the longer run, and greater passive protection for offensive systems -- while placing less emphasis on booster proliferation. Alternatively, the Soviets could emphasize options c) through e) -- defense for offensive systems -- while achieving cost-savings by foregoing the short-run, ground-based ABM solution.

THE DECISION TIME FRAME

Development, testing, and deployment of space-based laser BMD by the United States will require a sustained effort over a number of years. The Soviets will therefore have considerable opportunity to observe the pace and direction of the U.S. effort as they design their response. There will, however, be certain "benchmark" periods during which the pressure for decision will be more intense. Some of these relate to the timing of the United States' own BMD decisions and programs. For the purposes of this analysis, these critical periods are assumed to be 1982, the year of the second ABM Treaty review; 1983, the presumed year in which a major space-based laser BMD program is initiated; 1987, when the third ABM Treaty review occurs, and the first test bed is established; and the mid-to-late 1990s, when space-based laser BMD systems reach full deployment.

During the 1982 ABM Treaty review the United States did not insist upon renegotiation of or withdrawal from the ABM Treaty of 1972, although some dissatisfaction with the current situation may have been expressed to lay the basis for a tougher stance in the 1987 review. The Soviet Union therefore was not faced with the need to make fundamental decisions on BMD in 1982. In 1983, however, the increased commitment to BMD which a laser program initiation would signify might be interpreted in Moscow as "the handwriting on the wall." Important decisions on the future direction and scale of Soviet BMD might therefore be forthcoming in 1983 or early 1984. By 1987, the establishment of a test bed for U.S. space-based laser BMD would certainly call for a Soviet response if one had not already been made, and the ABM Treaty review scheduled for that year would provide a convenient forum for airing the Soviet political and strategic concerns.

But the Soviet Union does not simply react to U.S. initiatives. Important changes in Soviet strategic priorities and programs must be decided upon within the bureaucratic structure of Soviet defense planning. This system is dominated by the five-year planning cycle, although important flexibility is offered through the institution of annual plans. While no precise information on Soviet defense planning schedules is available in open sources, a former Polish military economist, Michael Checinski, has recently provided an "inside" view of the Polish planning process and suggests that the Soviet system, on which the Polish planning cycle is based, is probably very similar.[56]

According to Checinski, the process of preparing the new defense Five Year Supply Plan in Poland (and presumably in the U.S.S.R. as well) begins in September, sixteen months prior to the beginning of the subsequent Five Year Plan (FYP) period. During this time, the General Staff sends directives to the supply departments of the Ministry of Defense for initial feedback on capabilities. Over the following Spring, the Military Group of Gosplan integrates requirements into a draft plan. The Military Industrial Commission then coordinates the plan-proposal with its views of operational technical and national economic capabilities, and submits its revised version to the General Staff. After the General Staff and Minister of Defense have made their final amendments in late Autumn, the plan-proposal is submitted for final discussion to the Politburo-level Defense Committee. Formal approval of the document by this group usually comes in December, one month prior to beginning of the FYP period. After it is adopted, the plan-proposal is given the title "Supply Plan of the Armed Forces" and is sent for implementation to all levels of the military-industrial complex.[57]

The Soviet Union is currently in the midst of the Eleventh Five Year Plan (1981-1985). The Twelfth FYP will run from 1986-1990, and the Thirteenth FYP will cover the years 1991-1995. These three planning periods therefore encompass the critical time frame for U.S. and Soviet BMD developments. By applying Checinski's planning process model to these three Five Year Plans and superimposing its schedule on a time line depicting a possible development schedule for U.S. space-based laser BMD, one can obtain a projection of critical Soviet decision-making periods (see Figure 3). By the time planning for the Twelfth FYP begins in late 1984, the Soviet General Staff will have had about a year to observe the initial scope, pace, and success of U.S. space-based laser BMD research and development. If Moscow is convinced that the U.S. Government is seriously committed to this program, a decision might be made in later 1984-1985 to incorporate into the Twelfth FYP a program for an increased laser BMD research and development effort. Conceivably, the U.S.S.R. could also preempt the 1987 ABM Treaty review at this point by pursuing an early nationwide expansion of ground-based ABM. It is more likely, though, that Moscow would try to avoid the onus of being "first to withdraw." The Soviets might therefore try to derail the U.S. program through diplomatic means during the 1983-1986 period, while simultaneously pursuing a vigorous R&D effort in laser BMD themselves as a technological hedge.

By 1987, the U.S. program should be advancing at a steady pace, and the Soviets could hope to gain political capital by using the Treaty review of that year as a forum to place the blame for "a new round in the arms race" upon the United States. It is at this point that the Soviet leadership might choose to deploy a nationwide ground-based ABM system or to proliferate ICBM boosters beyond SALT II constraints. Since such a decision would occur in the middle of a five year plan period, though, it would not be easily made. Funds would have to be diverted in the annual plans from programs already budgeted in the Twelfth FYP. This would not pose an insurmountable obstacle. However rare, there are precedents for abrupt shifts in Soviet program budgeting. The termination of the SS-16 ICBM at an advanced stage of development is a case in point.[58] Still, anticipation of the adjustment problems which a major inter-plan shift in priorities in the context of a taut planning system would entail could lead the Soviet Union to make its basic decision earlier, in the 1984-1985 planning

FIGURE 3. CRITICAL PERIODS FOR SOVIET BMD DECISION-MAKING

Shaded areas represent critical time periods.

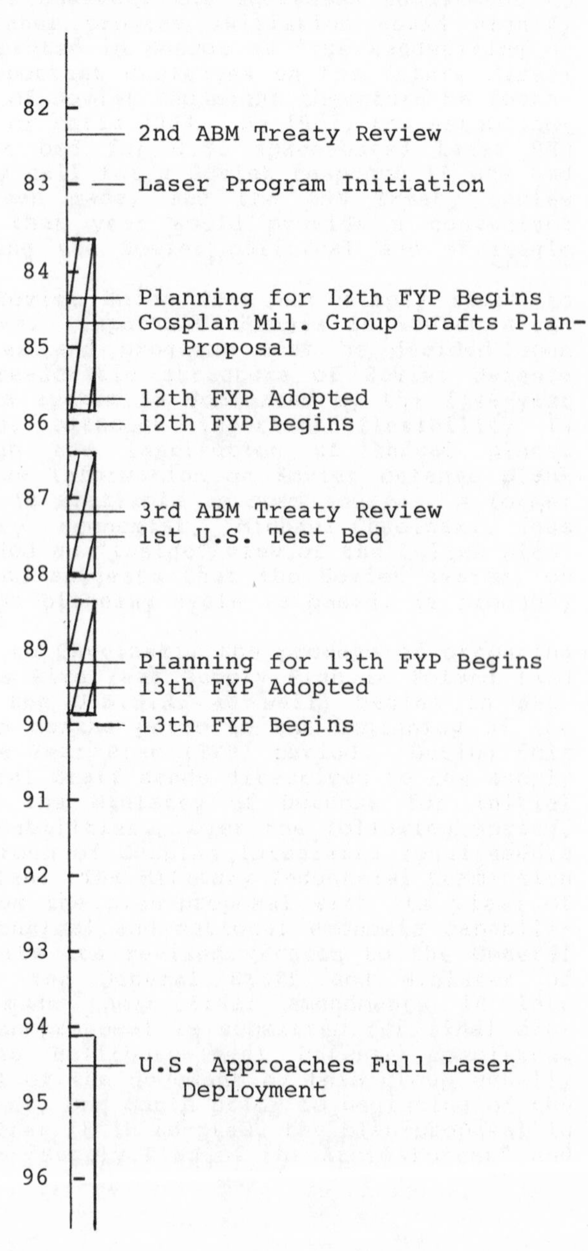

period for the Twelfth FYP. In any case, preliminary programs would probably be further enhanced in budgetary terms during the Thirteenth FYP, as the U.S. laser program begins to mature. Assuming that the Soviets have previously opted for a short-run ground-based ABM approach, the Thirteenth FYP might see the gradual phase-out of that program in favor of full-scale space-based laser BMD development.

BMD POLICY AND SOVIET STRATEGY

Prior to the signing of the ABM Treaty in 1972, the U.S.S.R. maintained a firm strategic-doctrinal emphasis on the value of anti-ballistic missile defense. In 1967, for example, Major General of Engineering-Technical Services I. Anureyev wrote in the General Staff journal, Military Thought, that "a most important factor which makes it possible to accomplish the task of changing the correlation of forces in one's own favor is anti-air defense (anti-missile and anti-space)".[59] Premier Kosygin rejected any notion that BMD might be destabilizing in a Western news conference in 1967, saying:

> As regards an anti-missile system, our positions is well known. We believe that the discussions should center not on merely the problem of an anti-missile defense system. Because, after all, the anti-missile system is not a weapon of aggression, of attack; it is a defensive system. And we feel therefore that what should be considered is the entire complex of armaments and disarmament questions.
>
> Because, otherwise, if -- instead of building and deploying an anti-ballistic missile system -- the money is used to buildup offensive missile systems, mankind will not stand to gain anything. It will, on the contrary, face a still greater menace and will come still closer to war. And we therefore are in favor of considering the whole range of questions relating to arms and disarmament, and we're ready to discuss that question -- the general question of disarmament.[60]

Damage-limitation for the Soviet Union was considered an integral part of Soviet strategy, which viewed nuclear war as an unprecedented phenomenon in terms of the level of violence it would entail, but a conflict which nonetheless could be fought for politically meaningful ends and could be pursued to a victorious conclusion. As Kosygin's 1967 remark clearly indicates, the U.S.S.R. did not accept the prevailing Western view of the time that security in the nuclear age should be sought through a deterrence strategy based ultimately on the threat of a massive retaliatory blow (Mutual Assured Destruction). But by signing the SALT I ABM Treaty, which limited each side to two ABM sites (later reduced by a subsequent protocol to only one site), the Soviet Union accepted restrictions which severely constrain its ability to protect Soviet territory from enormous nuclear devastation. Some Western analysts believe this act signalled a fundamental break with past Soviet military attitudes. Raymond Garthoff, Executive Secretary of the U.S. SALT I delegation writes:

> In SALT, the Soviets have accepted mutual deterrence, both through advocacy of equal security for the U.S. and the U.S.S.R. and even more tellingly through sponsoring an ABM limitation specifically precluding a defense of the country against overwhelming strategic missile attack and thus ensuring mutual vulnerability.[61]

While mutual vulnerability may indeed be a technological fact at the present time, Garthoff's contention that the Soviet leadership views it as a desirable condition is open to serious question. Since the ABM Treaty was signed, the U.S.S.R. has continued to conduct R&D on a variety of BMD technologies. In addition, Soviet military spokesmen continue to speak of the need for ballistic missile defense and for a traditional concept of victory. In 1976, for example, Marshal of the Soviet Union G.V. Zimin, Chief of the Military Academy of the PVO Strany (National Air Defense Forces), wrote that "now victory or defeat in war has become dependent on how much the state is in a position to reliably defend the important objects on its territory from the destruction of strikes from air or space."[62] A portion of this damage-limitation task can be achieved through offensive preemption. As Lieutenant General D. I. Shuvyrin noted in 1968:

> One must keep in mind that the agressors will not be able to make full use for their purposes of their strategic means of attack. A portion of these means of weapons delivery

> will be destroyed or damaged before their launching while they are still on their launch sites, bases, and airfields; another portion will be destroyed or damaged by the weapons of the National Air Defense in flight at the approaches to the territory of the socialist camp; still another portion of the missiles and aircraft will fail to reach their targets for technical reasons.[63]

Despite the contribution of offensive systems to the damage-limitation mission, Soviet sources, even in the post-SALT period, continue to claim that,

> The enormous destructive power of nuclear warheads raises the necessity of destroying all targets without exception, which accomplished a breakthrough into the interior of the country from air or space.
>
> All of these conditions put before the air defense complex and responsible tasks, the resolution of which will be determined by the ability to repulse strikes not only of aerodynamic, but also of ballistic, means of attack.[64]

The U.S.S.R. currently lacks the capability to fulfill these ambitious goals. But the continued doctrinal emphasis on the need for BMD, coupled with ongoing R&D efforts, suggests that the Soviet military leadership retains a keen interest in the potential military advantages of reduced homeland vulnerability. No exact estimate of Soviet ruble expenditure on ABM efforts is available, but the Defense Intelligence Agency judges them to be "considerably greater" than those of the United States. Furthermore, while U.S. R&D programs are technology and component programs, Soviet efforts are geared toward achieving deployable weapon systems, an approach requiring considerably more cost for construction and testing.[65] "The Soviets right now are expending considerable resources and manpower on the research and development associated with new ABM systems," said DIA's Deputy Director for Scientific and Technical Intelligence in 1979. "I don't think they are lacking at present for resources or priority or commitment to their ABM developments."[66] The CIA does not have a "useful estimate" of Soviet spending specifically on laser research, but believes that RDT&E on laser weapons constitute only a small percentage of total Soviet spending for military RDT&E.[67] Nevertheless, a

Rand Corporation study reportedly concluded that lasers are now receiving the same high-level political support in the Soviet weapons development system that was given earlier to nuclear weapons and ICBMs.[68] It therefore appears that those BMD options which seem attractive on technological grounds are sufficiently consonant with Soviet strategic goals and investment priorities to gain the acceptance of the Soviet General Staff. The U.S.S.R. has a continuing theoretical commitment to active missile defense, one which has been constrained in practice only by technological limitations, not by a lack of appreciation of the benefits of damage-limitation.

Institutionally, PVO maintains a credible bargaining position in the internal battle for scarce resources. The service's commander, Marshal of Aviation Aleksandr Koldunov, is also a Deputy Minister of Defense, and in 1981 was promoted from candidate membership of the CPSU Central Committee to full membership.[69] The PVO is the second largest of the five Soviet services. With some 550,000 men, it almost equals the manpower staffing of the entire U.S. Air Force.[70] A comparison of PVO's share of the Soviet military budget is provided in the following table.

Percent Share of Operation and Investment Spending (based on 1970 rubles)[71]

	1967	1970	1973	1977
Strategic Rocket Forces	10	7	5	8
Ground Forces	21	22	22	22
PVO Strany	14	15	12	12
Air Force	17	19	26	22
Navy	22	22	19	20

While the PVO's share declined as 1960s-vintage anti-air and anti-ballistic missile systems became fully deployed, so too did the percentage shares of the SRF and Navy. The Ground Forces held fairly steady, and only the Air Force showed marked growth. The PVO, however, is projected to show large budgetary gains both in absolute and relative terms, as new systems come on line. According to CIA, "The most significant trend in the distribution of defense spending through the mid-1980s is likely to be in strategic defense..."[72] This increase will be due primarily to the procurement of a new generation of surface-to-air missiles (SAMs) for improved air defense against low-altitude bomber and cruise missile attack, plus the procurement of a new interceptor aircraft. Unless immediate priorities change, ABM development will not account for the major share of PVO's projected growth; nevertheless, the air defense force's active procurement programs indicate its institutional strength. To the extent that budgetary clout implies strategic planning influence, the Voiska PVO can be expected to enjoy an increasingly important say in military planning.

With BMD technology becoming more sophisticated and effective, PVO can also expect to regain some of the status it apparently lost in the wake of the 1972 ABM Treaty. The mission of anti-rocket defense (PRO) may once again appear feasible to Soviet planners, and PVO would thus gain important strategic leverage in fighting for greater BMD funding.

An effective Soviet BMD capability would force the United States to undertake a major reassessment of the strategic balance. A credible Soviet space-based laser BMD weapon system would immeasurably complicate U.S. retaliatory calculations while buttressing Soviet prospects for war survival and post-attack recovery. In addition, by denying the United States easy access to its primary target set -- Soviet military forces and command and control assets -- Soviet laser BMD would undermine U.S. efforts to obtain greater time-urgent counterforce capability through improved accuracy and enhanced ICBM and SLBM hard-target kill capability.

With this strategic promise on the technological horizon, Soviet compliance with the 1972 ABM Treaty could well become problematic *whether or not* the United States seeks a change in the *current BMD status quo*. In the absence of Soviet BMD deployment, the U.S.S.R. may find the survivability of its own time-urgent, hard-target capable ICBM force threatened. The pressure to reconstitute a secure and accurate second-strike ICBM capability through BMD deployment would be especially strong if Moscow believed it could attain this advantage well before the U.S. could attain a comparable capability. For this reason, a substantial U.S. research effort in the area of laser BMD would be prudent as a hedge against and deterrent to an attempted Soviet "break-out" from the ABM Treaty, regardless of whether or not the U.S. ultimately chooses to develop and deploy such a system.

Although Soviet military writings and strategic R&D indicate a consistent interest in damage-limitation for an improved war-outcome in the event deterrence fails, recent statements by some Soviet political leaders, including the late Leonid Brezhnev himself, have been interpreted by some Western analysts as signalling a concensus within at least the civilian Party elite that no one, including the Soviet Union, could "win" a nuclear war. In his report to the Twenty-sixth Party Congress in February 1981, Brezhnev stated, "To try to prevail over the other side in the arms race or to count on victory in a nuclear war is dangerous madness".[73] Though such statements are still relatively few in number, they have appeared with increasing frequency since about 1977. Clearly they present a different tone from statements found in much of the military press. Those Western analysts who conclude from these pronouncements that the Soviet leadership no longer believes victory in a nuclear war to be possible under any circumstances infer that the U.S.S.R. must eventually lose interest in BMD for the purpose of damage-limitation. This notion clearly challenges the views set forth in this paper that (1) Soviet BMD research and the possibility of an attempted ABM "break-out" in the absence of a comparable U.S. program should be taken seriously, and (2) that deployment by the United States of an effective BMD system would greatly enhance the United States' ability to deter the Soviet Union. There is considerable evidence, however, that these skeptical views do not constitute a fundamental break with "traditional" Soviet military doctrine. Rather, they appear either to be part of a concerted propaganda effort aimed at dispelling Western alarm over the Soviet strategic buildup (thereby to forestall Western efforts to redress the balance), or to be the views of certain individuals within the leadership who, despite their influence, have not as yet fully articulated their opinions and have been unable or unwilling to bring about major alterations in Soviet operational strategy.

In the first place, the Brezhnev statement does not explicitly contend that victory in a nuclear war is impossible, only that one cannot count on it. In other words, given the uncertainties involved in strategic nuclear calculations, overconfidence would be a supreme folly. This was a sensible, and encouraging, view coming from the top Soviet leader. It in no way implies, however, that, should deterrence fail, the Soviets would not seek to wage nuclear war in a manner designed to optimize Soviet chances for a favorable war outcome -- i.e., one in which the attainment of meaningful political goals is facilitated. Secondly, Brezhnev said only that one cannot expect to prevail in the arms race, not that one cannot prevail in nuclear war. The Brezhnev statement itself is thus ambiguous. Moreover, it is not supported by more detailed discussions in the military or political press of the operational significance and force structure implications of the new position, if indeed there has been any substantive change at all. Yet surely a revision of strategic doctrine of this magnitude would require operational doctrinal reforms. In addition, investment in BMD R&D should decline. But to date none of these signals have been seen. Perhaps they will eventually appear, but until they do, it is difficult to place much stress on statements such as Brezhnev's. Consequently, so long as Soviet strategic "reforms" remain largely undefined and undeveloped, the arguments presented in favor of U.S. BMD deployment remain in full force.

But what of the ABM Treaty? Did it or did it not constitute Soviet abandonment of damage-limitation in favor of a strategy of mutual assured destruction? The evidence at hand suggests that Soviet acceptance of ABM limitations represented a short-run effort to compensate for technological inferiority, not an abandonment of a damage-limiting, warfighting strategy.

THE SOVIET APPROACH TO TECHNOLOGICAL CHALLENGE: CASE STUDIES

Some insight into the likely Soviet response to U.S. decisions can be gained by examining Soviet behavior in the past when the U.S.S.R. has been presented with technological challenges in weapon development. Three cases seem most relevant -- the Soviet Union's effort to limit ABM deployments in SALT I, its response to the PRC's ICBM program, and its campaign against the deployment of the enhanced radiation warhead in Europe.

A. The Role of the ABM in SALT I: Meeting Technological Challenge Through Arms Control

Although Khrushchev boasted in 1962 that Soviet ABMs could "hit a fly in outer space", it was less certain that they could strike a real target. The weapon, code named *Griffon* by NATO, was a two-staged, liquid-fuel missile with an effective zone of altitude of only twenty miles and with a warhead in the kiloton-range. This, however, was only a start. In November 1964, a new ABM, the *Galosh*, was paraded in Red Square, revealing a considerably improved ABM technology. *Galosh* was a solid-fuel missile with a slant range of 100-200 miles, and carried a 1-2 megaton (MT) warhead. The technology so impressed the United States that when the idea of SALT was first broached to the Soviets in December 1966, the American proposal envisaged only an ABM limitation. The Soviets, on the other hand, replied that such talks would be impossible unless they encompassed offensive strategic weapons as well.

In perspective, this circumstance seems ironic, because four years later the situation would be just the reverse -- the U.S. wanting offensive weapons treated in conjunction with defensive systems, and the Soviet Union seeking to decouple the two. But in January 1967, when the Soviets first replied to the Johnson Administration's initiative, they did not wish to restrict their freedom of action in the one area in which they appeared to hold the advantage. At the time, Galosh was already deployed around Moscow and Griffon around Leningrad, whereas the U.S. ABM program was not yet operational.[74] Moreover, an ABM curtailment in 1967 without an offensive cutback by the United States would have left the Soviets in a grossly inferior position, for the United States at that time had preponderant offensive superiority.

By mid-1967, however, Soviet confidence in its ABM program had considerably waned, apparently due to technical difficulties, rising costs, and system inefficiencies. Moreover, Galosh stood to be seriously degraded by the American technique then under research of equipping land- and sea-based ballistic missiles with multiple independently targeted reentry vehicles (MIRVs).[75] The Soviets' skepticism concerning their ability to surmount these difficulties is evident in a June 1967 article in Voyennaya mysl' in which Major General I. Anureyev created a model of the correlation of forces which assumed the probabilities of overcoming the anti-missile defenses of both sides to be unity, i.e., the probability of a successfully ballistic missile defense was assumed to be nil. While Anureyev added the disclaimer that his probability coefficients were intended only for the purposes of illustration, others in the model were quite realistic, and it seems likely that he did not consider the one for ballistic missile defense to be farfetched.[76] Three years later, while SALT I was in progress, A.A. Sidorenko, author of The Offensive, would manifest the same pessimism by listing as one of the main qualities of nuclear missiles their "invulnerability in flight".[77]

Tangled in increasing difficulties, the Soviets in 1967 curtailed their ABM deployment by 33 percent. Still, they did not abandon the idea of ballistic missile defense. Sixty-four of the originally planned ninety-six ABM launchers were deployed around Moscow, R&D was kept at a high level, and three large, sophisticated, phased-array radars were constructed.[78]

To complicate matters still more, the Johnson Administration determined in September 1967 to deploy its own ABM system. Though Sentinel was to be only a thin defense, the Soviet Union feared it might be enlarged into just the sort of system that President Nixon was later to propose under the rubric Safeguard.[79] A thick, silo-protecting ABM force would have foiled the Soviets' first-strike capability unless they could have developed sophisticated penetration aids.[80] For this reason, the Soviets feared America's now more advanced ABM technology, which may have been some ten years ahead of the Soviets' by 1969, when SALT began.[81] It is thus not a coincidence that the Soviet Government finally agreed to "an exchange of opinion on arms limitation, including anti-missile systems" on June 27, 1968, only three days after Congress decided to fund Sentinel.[82] From that time on, the Soviets would seek to use SALT to retard the American ABM program, repeatedly threatening that deployment of Sentinel or Safeguard would start a new round in the arms race and seriously inhibit the conclusion of a SALT Treaty.[83]

From the earliest negotiating sessions, the Soviet Union supported a low ceiling on ABM systems, accepting the U.S. proposal that each side be limited to one ABM site, to be located around its National Command Authority (NCA), i.e., the national capital.[84] This would negate the United States' technolgoical superiority by allowing the Soviet Union to overwhelm the American defense by means of a large-scale ballistic missile attack. That the United States could do the same to the Soviet Union made little difference, for the Soviets had, by the time SALT began, determined that their ABM system could not stop U.S. MIRVed missiles, anyway. Thus, when the United States, alarmed over the Soviet Union's growing hard-target kill capacity, and seeking therefore to disengage itself from its previous ABM stance, tried on March 26, 1971, to salvage the Safeguard program by proposing that the U.S. be allowed four ABM sites arounds its ICBMs while the U.S.S.R. remained restricted to the single ABM site already in place around Moscow, the Soviet delegation replied with an angry "no." Subsequent efforts by the United States to retain some meaningful ABM capacity were also rejected, and, in the end, the U.S. essentially acceded to the Soviet position by settling on two ABM sites per side, one around the NCA and one around an ICBM installation. The provision for one ICBM defense site was but a minor concession by the U.S.S.R. since the strict limit of only 100 ABMs per site assured that they would not pose a significant threat to incoming Soviet ICBMs.

On the other hand, a low ceiling, rather than an outright ban, would at least provide some room for the concept of ballistic missile defense. This was important to the Soviet Union for two reasons. First, even a limited ABM deployment would suffice against a third-power threat -- particularly from primitive Chinese missiles.[85] Second, the Soviet Union had never abandoned in principle the idea of anti-missile defense. As later developments in Soviet satellite interceptor, surface-to-air missile (SAM), and civil defense programs were to demonstrate, the Soviet Union's consent to limit ABM systems was motivated by technological setbacks, not by any shift in Soviet strategy away from the commitment to homeland defense.

If such a defense were achieved, it would provide immense strategic advantages. Soviet generals therefore warned against assuming that no technology could defeat the ICBM, arguing that "no matter how the weapons arsenal changes qualitatively, it is impossible to absolutize one type of weapon, even if it is the most powerful."[86] The Soviets felt it would be unwise to scrap the ABM program due to current ineffectiveness because experience has shown that weapons once considered obsolescent have later proven useful in improved models or in different roles.

The Soviet Union's difficulties with its ABM systems lent added weight to its submarine force. Submarine-launched ballistic missiles (SLBMs) could have three missions in Soviet strategy: to deliver a short warning first-strike on Strategic Air Command bases and to "pin down" American ICBMs,[87] to provide a secure reserve force for conducting war after the initial nuclear exchange (or for interwar bargaining), and to provide a hedge against future U.S. counterforce capabilities. The import of the latter mission was enhanced by the deficiencies of the *Galosh* system. Consequently, the Soviet Union undertook a major SLBM building program in 1968 and sustained it throughout the course of SALT.[88] At the outset of the negotiations, Kissinger had instructed the American delegation to accede to the Soviet desire not to include SLBMs in the agreement. But when this decision was belatedly communicated to the U.S. Government in the Summer of 1971, Kissinger discovered to his consternation that all relevant agencies strongly desired that an SLBM limitation be included -- this, no doubt, in an effort to restrain the Soviet buildup which was threatening to undo U.S. numerical superiority. Not until April 1972 during his trip to Moscow did Kissinger manage to persuade the Soviets to include launchers on nuclear-power submarines (SSBNs) -- the Soviets were adamant that those on diesel submarines by excluded -- and then only by agreeing to ceilings on both SSBNs and SLBM launchers far in excess of what either the Soviet Union or the United States had at the time. Indeed, the ceiling was set at the maximum level that U.S. intelligence believed the U.S.S.R. could have reached within a five-year period had there been no agreement.[89] In effect, therefore, no real restraint was placed on the Soviet SSBN and SLBM buildup, precisely the condition the Soviets had desired when they entered SALT.

The ABM Treaty also had an important ramification on the disposition of Soviet "heavy" (counterforce capable) missiles in the Interim Offensive Agreement. Here the impact had less to do with the ABM limitation _per se_ than with the manner in which the Treaty was _signed_. It will be recalled that the Soviet Union had originially insisted that any ABM Treaty also deal with offensive strategic weapons. But as the Soviets' own offensive capabilities strengthened and their ABM capabilities declined, they reversed their position radically by proposing in December 1970 that the initial SALT agreement limit only ABM systems. This was unacceptable to the United States, because its focus had also shifted from concern over Soviet ABMs to alarm over the U.S.S.R.'s deployment of the SS-9 ICBM, which was capable of delivering a 20 MT warhead. Such a weapon could be designed for one thing only: to destroy missile C^2 assets and ICBMs in hardened silos, thereby denying the United States the secure second-strike triad on which its nuclear strategy rested. Finally, after a deadlock of almost a year, the two governments revealed, on May 20, 1971, that secret, parallel negotiations which had been conducted by President Nixon and Premier Kosygin on the one hand, and presidential advisor Kissinger and Ambassador Dobrynin on the other, had produced a "breakthrough" toward an agreement. The SALT delegations would henceforth seek to work out two separate documents: an ABM Treaty, and an "interim agreement" on "certain measures with respect to the limitations of offensive strategic weapons". At first glance, it appeared the United States had won; offensive weapons would be dealt with in the agreements. But the decoupling of the offensive and defensive issues carried with it two effects which ultimately transformed the Interim Offensive Agreement into a largely cosmetic document. First, the IOA, being merely a temporary arrangement on "certain measures," lacked the stringency and definition of the ABM Limitation Treaty, and therefore contained more loopholes for further offensive weapon development. Second, the decoupling undercut the United States' bargaining strategy on the crucial matter of Soviet heavy missiles. The United States had hoped to use the Safeguard ABM system as a bargaining chip with which to obtain Soviet concessions on heavy missile deployment. But the separation of the two issues made this no longer possible. Any attempt by the American delegation to link the two was open to the charge that the United States was confusing the issue and trying to renege on matters already settled. This subtle negotiating coup marked the end of U.S. chances to obtain significant restrictions on Soviet counterforce-capable missile deployments during the tenure of SALT I.

In and of themselves, the SALT I Accords did not give the Soviet Union a counterforce capacity. In 1972, an attack expending 75 percent of the Soviet Union's available throwweight would have left the United States with some 6 million pounds of surviving throwweight of its own. This would have been sufficient to enable the U.S. to eliminate most of the U.S.S.R.'s remaining forces. The Soviets could hope to emerge from the American retaliatory strike with only about 0.5 million pounds of throwweight, too little to provide a politically decisive advantage.

The value of SALT to the Soviets, however, was that it did not restrict further expansion of their nuclear forces. Thus, within the framework of the SALT I Treaties, the Soviet Union by 1974 had already so increased its counterforce capabilities that a U.S. nuclear retaliation would be unable to reduce the U.S.S.R.'s reserve throwweight to below 2 million pounds.[90] The prospect of this much throwweight being left for a third strike would be so threatening a scenario to the United States that it might consider suing for peace without retaliating with a second strike against the U.S.S.R.

The major benefit of the SALT I Accords from the Soviet perspective, then, was that the Treaties effectively annulled the American lead in ABM technology while simultaneously allowing great latitude under the IOA for increased offensive strategic power. Since the Treaties were signed in 1972, the U.S.S.R. has availed itself of this opportunity by deploying new heavy missiles and by incorporating MIRV technology in many of its strategic systems. The extent to which these force improvements can be offset by new U.S. weapon systems -- especially the Trident submarine, the cruise missile, and (perhaps in the future) the MX in a survivable mode -- remains a matter of debate. But one point seems clear: SALT does not, in the Soviet conception, mean the end of strategic competition. Rather, it channels that competition in particular directions, and the Soviets endeavor to assure that this channelling occurs in directions in which it feels most able to compete.

If this interpretation of the motives behind Soviet accession to the SALT I ABM Treaty is correct, one could expect that future Soviet advances in BMD technology could lead the U.S.S.R. to reconsider the Treaty -- either by relaxing the Treaty's restrictions through renegotiation, or by outright withdrawal from the agreement. The continued viability of the 1972 accord is not dependent solely upon the United States. Soviet long-term interest in strategic defense, even in the post-1972 era of deterrence, is well expressed by Col. Bondarenko:

> If potential opponents possess weapons of mutual destruction, decisive advantage goes to that side which first manages to create a defense from it. The history of military arms development is full of examples in which weapons which seemed irresistible and frightening are, after some time, opposed by a sufficiently reliable means of defense. Thus, an absolute limit to the development of military power from the position of the internal development of military affairs and military-technical progress cannot exist.[91]

Secondly, the ABM case study suggests that U.S. development of a superior BMD technology could bring the Soviet Union to the negotiating table for serious arms control talks. A favorable outcome from such negotiations will require careful and skillful planning on the part of U.S. policy-makers, but there is no reason why BMD could not be used by the United States as leverage in efforts to obtain meaningful Soviet concessions on offensive and defensive issues.

B. The "Neutron Bomb": Meeting Technological Challenge Through Political Propaganda

When President Carter proposed the introduction of enhanced radiation warheads into Europe, the Soviet Union responded with a massive propaganda campaign directed primarily to Western Europe. Moscow's political offensive fostered three arguments against ER weapons: (1) that they complicate arms control efforts, (2) that they are strategically destabilizing, and (3) that they are designed to enable the United States to further its own interests at the expense of Europe. The broader thrust of the campaign, to which each of these points contributed, was to use the ER debate as a means of weakening NATO Europe's ties to the United States.

A typical example of Moscow's claim that ER weapons would be detrimental to prospects for arms control appeared in an article by Sergei Konstantinov in the journal New Times, a Soviet publication sold in a variety of languages throughout the globe. Kostantinov wrote:

> All sober-minded people realize that the U.S. Administration's deicison to start the production of neutron bombs and new types of missiles is opening a new round of the arms race at a time when favorable conditions have appeared for ending it. The neutron bomb decision is complicating the current talks on military detente.[92]

On the destabilizing effects of an ER weapon deployment, a Soviet academician claimed in the same journal that the neutron bomb "impedes efforts to prevent the spread of nuclear weapons, and increases the probability of accidental, unsanctioned outbreak of nuclear war".[93] The Soviet claim that the ER warhead would be inimical to NATO Europe's interests was graphically expressed in a New Times article by Lev Bezymensky, who wrote that

> the new weapon is a new variant of the old American policy with regard to NATO which boils down to the primitive but unambiguous formula: "Let someone else do the dying." This prospect frightened Western Europe at the time of John Forster Dulles' brinksmanship. It is still less attractive today when tangible vistas of peaceful and mutually beneficial coexistence are opening before Europe.[94]

These themes might well be sounded again in a Soviet political campaign against U.S. BMD deployment. Although the Soviets have never claimed in their military press that BMD is destabilizing, they will not hesitate in their international propaganda to echo the concerns of those in the United States who view it as such. Indeed, a recent statement from the U.S.S.R.'s Institute for World Economics and International Relations exemplifies this tactic:

> Many foreign specialists justifiably point to the destabilizing effect on the global strategic situation of the creation and deployment in space of effective combat systems designed to destroy satellites and ballistic rockets in flight, and to put out of action [ground-based] means of anti-air [PVO] and anti-space [PRO] defense. Such space systems cannot be viewed as other than part of an effort to obtain a "first-strike" potential.[95]

It is important to note that this statement accuses only space-based BMD of being destabilizing. Ground-based PRO (anti-space defense) is explicitly exempted from this charge; and it is, of course, only the Soviet Union that currently maintains a deployed, ground-based BMD system.

Similarly, the Soviets would probably attempt to shake Alliance unity by implying that U.S. BMD would weaken America's commitment to European security. Moreover, the Soviets would charge that insofar as BMD increased the relative security of CONUS from nuclear destruction, nuclear war in Europe would be viewed in Washington as a "safe" option. This in turn would supposedly render the outbreak of nuclear war more likely, because U.S. leaders would have less fear of nuclear escalation. Though the United States might attempt to undercut this critique by offering a parallel anti-theater ballistic missile defense program to the Europeans as an appropriate means of countering the Soviet I/MRBM threat, many European leaders would still probably consider a U.S. deploy- ment of BMD to be a grave step backward in U.S.-Soviet relations.

Finally, if the Soviet Union refused to renegotiate the ABM Treaty and the United States then chose to withdraw from the Treaty unilaterally, the Soviets would attempt to win a diplomatic victory by charging that the United States had ruined all prospects for arms control. U.S. withdrawal from the only SALT Treaty still in force would be labelled "the death of arms control," and might be used by Moscow as an excuse not only for increasing its own BMD activities, but for enlarging its offensive strategic missile forces as well. As suggested earlier, economic constraints might retard this effort. Still, it is a serious possibility. The Soviets might choose to let the U.S. withdraw from the Treaty even if they were themselves already intent on deploying BMD systems proscribed by the agreement, in order to gain technological latitude while letting the United States bear the political costs. In any event, the Soviet Union will not make U.S. space-based laser BMD deployment politically easy. A major propaganda campaign appears already to be underway, and it can be expected to intensify over the coming months. Thus, the United States must be prepared to meet not only Soviet technological responses to U.S. laser BMD, but Soviet political responses as well.

C. The Chinese ICBM: Meeting Technological Challenge Through Diplomacy

The fundamental lesson to be learned from the Soviet Union's response to the PRC's ICBM program is that it has been diplomatic, not military. The development of a Chinese ICBM force represents a serious long-term threat to the U.S.S.R. At the present time, however, the Soviet Union could probably preemptively destroy Chinese nuclear forces with little difficulty. Nevertheless, Moscow has refrained from doing so, opting instead to use diplomacy to isolate the PRC and to reduce the value to the United States of the "China card." One goal of this policy is to reduce the likelihood and magnitude of military-related technology transfers from the West to the PRC. The second goal is to convince the United States that it must come to terms with the Soviet Union because China has little to offer.

The U.S.S.R. may briefly have contemplated a preemptive strike against Chinese nuclear forces in the aftermath of the Sino-Soviet border clashes of 1969. On August 18, 1969, a Soviet official in Washington asked a mid-level State Department official what the U.S. reaction to a Soviet attack on Chinese nuclear facilities would be. Rumors that Moscow was feeling out West European Communist Party leaders on the same subject persisted through mid-September. But Soviet hints at such activity suddenly disappeared, and on October 7, Moscow announced that border talks with the Chinese would begin later that month.[97] Perhaps the Soviets never considered preemption of Chinese nuclear forces seriously, but used the threat to bring the PRC to the negotiating table. But even if this option was seriously contemplated in the Kremlin, it was soon abandoned -- and abandoned <u>before</u> the United States responded to Soviet feelers. Thus, it was rejected by the Soviets on their own, without any pressure from Washington.

The implications of this incident for Soviet reactions to U.S. laser BMD deployment are significant. That the Soviets have not responded to an evolving Chinese technological challenge through direct military preemption suggests that a peacetime attack on U.S. space-based laser assets is unlikely. The situations are analogous, and if the option has not appeared desirable to Moscow with regard to the PRC, it is much less likely to appear so in relation to the United States, where possible retaliation would carry much more dangerous consequences. That the U.S.S.R. _may_ have contemplated preemption of PRC nuclear facilities in the heat of a political and low-level military conflict suggests that an attack on U.S. space-based laser weapons might be forthcoming should the Soviet leadership become convinced that war with the United States was immediately in prospect.

CONCLUSIONS

The lessons which stand out most clearly from these studies of Soviet BMD R&D, strategic doctrine, and historical behavior are that the U.S.S.R. seeks to overcome unfavorable military developments abroad not only through analogous research and development, but also through political-diplomatic measures, and that there is little evidence that the Soviet Union starts wars over issues of technology -- for the Soviet Union, technology, like war, is a tool of policy; policy and war do not serve technology. Historical precedent suggests that the Soviets would accompany a technological response to a U.S. laser BMD with a vigorous political campaign, a principal aim of which would be to drive a wedge between NATO Europe and the United States by depicting the U.S. as a warmonger responsible for the "death of arms control" and intent upon securing "Fortress America" at the expense of European security. Nevertheless, historical precedent also suggests that steadfast U.S. policy in the face of this propaganda campaign can provide sufficient leverage to bring the U.S.S.R. to the negotiating table. The _Safeguard_ incident in particular indicates that a vigorous U.S. space-based laser BMD program could induce the U.S.S.R. to negotiate seriously with the United States on such issues as total U.S.-Soviet missile payload disparities, numbers of heavy missiles, and other areas of concern to U.S. strategic planners. Finally, these case studies strongly suggest that, short of a prior Soviet decision to engage the United States in large-scale, and possibly nuclear, war, the U.S.S.R. would consider a direct attack against a deployed U.S. space-based laser BMD to be too risky for serious policy.

In terms of weapon procurement, the Soviet response to U.S. space-based laser BMD would probably combine active and passive measures, including comparable BMD deployments and possible increased numbers of offensive systems. The economic difficulty of pursuing large-scale offensive and defensive deployments simultaneously might, however, be so severe as to force greater reliance in the long run on one option (probably BMD) at the expense of the other.

NOTES

1. *Treaty Between the United States of America and the Union of Soviet Socialist Republics on the Limitation of Anti-Ballistic Missile Systems* (1972), Article V, Paragraph 1.

2. See "Go-Ahead on USAF's ASAT Program," *Air Force Magazine*, Vol. 64, No. 10 (Oct. 1981), p. 16; Edgar Ulsamer, "Soviet Military Power," *Air Force Magazine*, Vol. 64, No. 12 (Dec. 1981), pp. 48-49; and Richard L. Garwin, "Are We on the Verge of an Arms Race in Space?" *The Bulletin of the Atomic Scientists*, Vol. 37, No. 5 (May 1981), p. 50.

3. S. Stashevskii and G. Stakh, "Kosmos dolzhen byt' mirnym" [Space Should be Peaceful], *Mirovaia elonomika i mezhdunarodnye otnosheniia* [World Economics and International Relations], No. 2 (Feb.) 1982, pg. 18.

4. Ibid, p. 19.

5. See *infra*, pp. 106, ff.

6. *Soviet Military Strategy*, ed. by V.D. Sokolovskii, trans. and ed. by Harriet Fast Scott, 3rd ed. (New York: Crane, Russak, and Company, 1975: Russian edition, Moscow, 1968), p. 15.

7. Colonel V.M. Bondarenko, *Sovremennaia nauka i razvitie voennogo dela* [Modern Science and the Development of Military Affairs] (Moscow, 1976), p. 92.

8. M.M. Kir'ian, "Vnezapnost'" [Surprise], *Sovetskaia voennaia entsiklopediia* [Soviet Military Encyclopedia], ed. by Marshal of the Soviet Union A.A. Grechko (Moscow, 1976), p. 161.

9. Ibid.

10. For a further discussion of the role of damage-limitation in Soviet strategy, see Colonel William J. Barlow, "Soviet Damage-Denial: Strategy, Systems, SALT, and Solution," *Air University Review*, XXXII, 6 (Sept.-Oct. 1981), pp. 2-20.

11. Synergistic advantages deriving from the combination of active and passive measures which comprise the Soviet Union's overall strategic force posture are postulated, but not demonstrated, in J.W. Russel and E.N. York, Expedient Industrial Protection Against Nuclear Attack (Seattle, Washington: The Boeing Company, March 1980), p. 56.

12. "Strategicheskaia tsel'" [Strategic Goal], Sovetskaia Voennaia Entsiklopediia, Vol. 7 (Moscow, 1979), p. 552.

13. Marshal I. Kh. Bagramian, Istoriia voin i voennogo iskusstva [The History of War and Military Art], p. 489, cited in Leon Goure, Foy D. Kohler, and Mose L. Harvey, The Role of Nuclear Forces in Current Soviet Strategy (University of Miami, Center for Advanced International Studies, 1974), p. 108. 14. Krasnaia zvezda [Red Star], Oct. 30, 1970.

15. The first experimental test of a large thermonuclear explosion was successfully conducted by the United States on November 1, 1952. The "Mike" shot, as the explosion was called, yielded an energy equivalent of 10 megatons of TNT, but as Herbert York attests, the device was "very far from being a practical deliverable weapon". Based on liquid deuterium, requiring refrigeration to below -250°C, the entire mechanism required a substantial laboratory building to house it on Aluglat Island in Eniwetok Atoll. See Herbert York, The Advisors: Oppenheimer, Teller, and the Superbomb (San Francisco: W.H.Freeman and Co., 1976; pp. 82-83). On August 8, 1953, the U.S.S.R. conducted its first successful thermonuclear explosion in a test commonly known in the West as "Joe 4". The device apparently used lithium deuteride rather than liquid deuterium as its explosive fuel, thereby facilitating a substantial reduction in the device's size. While the device was still large and was mounted on a tower rather than being air-dropped, it may have been small enough to be considered transportable. (See York, ibid; pp. 89-92.) Soviet writers therefore claim to have been the first to create a "superbomb". For example, academician Andrei Aleksandrov described

the event in the following manner: "The most complex calculations for different variants of the bomb were completed, and on 12 August 1953, under I.V. Kurchatov's leadership, a test of this monstrous superbomb was carried out. Now you no longer had to worry -- the U.S. still did not have such a weapon, although it was working vigorously to create one. Our scientists -- especially our theoretical physicists and experimenters, mathematicians, and designers -- had in this region established the superiority of Soviet science". (A.P. Aleksandrov, "Iadernaia fizika i razvitie atomnoi tekhniki v SSSR" [Nuclear Physics and the Development of Atomic Science in the U.S.S.R.], in Oktiabr' i nauchnyi progress [October and Scientific Progress], Vol. 1; Moscow, 1967; p. 201.) Nevertheless, the U.S. Joint Committee on Atomic Energy believed the device to be "nondeliverable". See Arnold Kramish, Atomic Energy in the Soviet Union; (Stanford: Stanford University Press, 1959; p. 125.) Not until November 23, 1955, did the Soviets actually drop a hydrogen bomb from an airplane. (Kramish, ibid.) The first deliverable hydrogen bomb built by the United States was tested (but not by an air-drop) in the Spring of 1954. (York, The Advisors, p. 85). But Soviet sources consistently imply that the bomb exploded in August 1953 could indeed have been delivered by an aircraft. In Scientific-Technical Progress and the Revolution in Military Affairs, V.M. Bondarenko writes: "In the United States a hydrogen device was detonated in 1954 [sic]. This was not yet a bomb, but namely a device, and a very heavy and cumbersome one. No aircraft of those times could have carried it. In the Soviet Union, the testing of the first thermonuclear bomb occurred in 1953, and we, thus, were ahead of the Americans in developing the most powerful and advanced type of modern weapon." (V.M. Bondarenko "Scientific- Technical Progress and Its Effect on the Development of Military Affairs," in Scientific- Technical Progress and the Revolution in Military Affairs, ed. by Col. Gen. N.A. Lomov; Moscow, 1973: trans. by the United States Air Force, Officers' Library Series; Washington, DC: GPO, n.d.; p. 34) On Soviet missile developments, see Ibid, p. 35.

16. Brezhnev interview, Pravda, 24 Dec. 1977, available in translation in Current Digest of the Soviet Press, XXIX, 51,p.4.

17. U.S. Senate, Committee on Armed Services, Subcommittee on General Procurement, Soviet Defense Expenditures and Related Programs. Hearings. 96th Congress, 1st and 2nd Sess., Nov. 1,8, 1979; Feb. 4, 1980 (Washington, DC: GPO, 1980), p. 75. On Galosh's characteristics, see The Military Balance: 1981/82 (London: International Institute for Strategic Studies, 1981).

18. John W.R. Taylor, "Gallery of Soviet Aerospace Weapons", Air Force Magazine, Vol. 65, No. 3 (March 1982), p. 109.

19. Barlow, "Soviet Damage-Denial", p. 9.

20. Clarence A. Robinson, Jr.: "Emphasis Grows on Nuclear Defense", Aviation Week and Space Technology, Vol 116, No. 10 (March 8, 1982), p. 27.

21. U.S. Senate, Committee on Armed Services, Subcommittee on General Procurement, Soviet Defense Expenditures and Related Programs. Hearings, p. 119.

22. Walter Pincus, "Soviets Believed to Have Problems With New Typhoon Missile", Washington Post, 18, Jan. 1982, p. 15.

23. Barlow,: Soviet Damage-Denial", p. 9. Emphasis added.

24. Jacquelyn Davis et al., SALT II and U.S.-Soviet Strategic Forces (Cambridge, MA: Institute for Foreign Policy Analysis, June 1979), p. 10; and Barlow, "Soviet Damage-Denial", p. 6.

25. Barlow, "Soviet Damage-Denial", p. 18, 28n.

26. Secretary of Defense Harold Brown, Department of Defense Annual Report, Fiscal Year 1982 (Washington, DC: Jan. 1981), p. 112.

27. Department of State, SALT ONE: Compliance, SALT TWO: Verification, Selected Documents No. 7 (Washington, DC: GPO, Feb. 1976), pp. 2-4.

28. "Specification: Soviet Missiles", Aviation Week and Space Technology, Vol. 116, No. 10 (March 8, 1982), pp. 146-147. Senator Jake Garn (R-Utah) reports the SA-5's interception altitude to be as high as 150,000 feet. See Senator John "Jake" Garn, "The Suppression of Information Concerning Soviet SALT Violations by the U.S. Government", Policy Review (Summer 1979), p. 26.

29. "Specifications: Soviet Missiles", p. 146.

30. See U.S. Senate, Committee on Armed Services, Subcommittee on General Procurement, Soviet Defense Expenditures and Related Programs. Hearings, p. 97. Neither Aviation Week and Space Technology nor The Military Balance suggests an ABM role for this missile. See Taylor, "Gallery of Soviet Aerospace Weapons", Air Force Magazine, Vol. 65, No. 3 (March 1982), p. 110.

31. Brown, Department of Defense Annual Report, Fiscal Year 1982, p. 45.

32. On the Soviet Union's incremental approach to weapon design, see Arthur J. Alexander, Decision-Making in Soviet Weapons Procurement, Adelphi Paper Nos. 147 and 148 (London: International Institute for Strategic Studies, Winter 1978/79).

33. U.S. Senate, Committee on Armed Services, Subcommittee on General Procurement, Soviet Defense Expenditures and Related Programs. Hearings, pp. 78 and 83. See also Soviet Military Power (Washington, DC: GPO, 1981), p. 75.

34. Ibid, p. 83. 35. William J. Perry, The FY 1981 Department of Defense Program for Research, Development, and Acquisition, Statement of Undersecretary of Defense Research and Engineering William J. Perry to the 96th Congress, 2nd session, 1 Feb. 1980, p. 1-3. Soviet Military Power, p. 76, states "The Soviet high energy laser program is three-to five times the U.S. level of effort and is tailored to the development of specific laser weapon systems".

36. Perry, FY 81 DoD Program for RD&A, p. 11-9.

37. Ibid.

38. Albert Wohlstetter, _Legends of the Strategic Arms Race, Part II: The Uncontrolled Upward Spiral_, United States Strategic Institute Report 75-1 (Washington, DC: United States Strategic Institute, 1975), p. 72; and Thomas W. Wolfe, _The SALT Experience_ (Cambridge, MA: Ballinger Publishing Co., 1979), pp. 118-122.

39. "Soviet Military Programs in Space Move Forward", _Aviation Week and Space Technology_, Vol. 116, No. 10 (March 8, 1982), p. 106 Emphasis added.

40. U.S. Senate, Committee on Armed Services, Subcommittee on General Procurement, _Soviet Defense Expenditures and Related Programs. Hearings_, pp. 68-69.

41. Clarence A. Robinson, Jr. "GAO Pushing Accelerated Laser Program:, _Aviation Week and Space Technology_, Vol. 116, No. 15 (April 12, 1982), p. 19.

42. Garwin, "Are We On the Verge of an Arms Race in Space?" p. 52; and Robinson, "GAO Pushing Accelerated Laser Program", p. 18.

43. According to the SALT II Treaty, any modifications of existing ICBM systems would be limited to a plus or minus five percent change in such parameters as length, diameter, throwweight and launch-weight except for certain specific exemptions regarding reductions in reentry vehicles, decoy warheads, and fuel. Soviet acceptance of this restriction indicates that their immediate plans for fourth-generation modifications would not substantially alter missile payloads. The U.S.S.R. originally sought a provision allowing considerable "downsizing" of missile parameters, in excess of five percent. This raised suspicisions in the U.S. that the Soviets might seek to introduce as "modifications" essentially new weapon systems, thereby circumventing SALT II's limitation of both sides to deployment of only one new ICBM type during the Treaty's tenure. The Soviet stance also seemed to imply that some fifth-generation systems under development might be significantly smaller than the SS-17, SS-18, and SS-19, due to improved propulsion and guidance systems. However, a concensus gradually emerged within the U.S. defense community that

the Soviet concern with "downsizing" was primarily related to missile testing practices, and did not necessarily imply much-reduced throw-weights for future Soviet missiles. See Strobe Talbott, Endgame: The Inside Story of SALT II (New York: Harper and Row Publishers, 1979), pp. 226-227, and 260-263.

44. Garwin, "Are We On the Verge of An Arms Race in Space?" p. 52.

45. Kosta Tsipsis, "Laser Weapons", Scientific American, Vol. 245, No. 6 (Dec. 1981), p. 57.

46. Under the Reagan Administration, the acronym SALT has been changed to START, for STrategic Arms Reduction Talks.

47. Central Intelligence Agency, National Foreign Assessment Center, Soviet and U.S. Defense Activities, 1970-1979: A Dollar Cost Comparison (SR80-10005, Jan. 1980), p.3.

48. Ibid., pp. 3-5.

49. Ibid., p.5.

50. Perry, The FY 1981 Department of Defense Program for Research, Development, and Acquisition pp. 1-1 to 1-2.

51. Central Intelligence Agency, Soviet and U.S. Defense Activities, 1970-1979: A Dollar Cost Comparison, p. 9.

52. Ibid.

53. U.S. Senate, Committee on Armed Services, Subcommittee on General Procurement, Soviet Defense Expenditures and Related Programs. Hearings, pp. 120-122.

54. Ibid.

55. Ibid.

56. Michael Checinski, A Comparison of the Polish and Soviet Armaments Decisionmaking Systems, R-2662-AF (Santa Monica: The Rand Corporation, Jan. 1981), p.v.

57. Ibid, pp. vi-vii, 19-23, and 76.

58. Perry, The FY 1981 Department of Defense Program for Research, Development and Acquisition, p. 11-8. Recently, doubts have arisen among defense analysis over whether the SS-16 program has really been cancelled.

59. Major General I. Anureyev, "Determining the Correlation of Forces in Terms of Nuclear Weapons:, Voyennaya mysl' [Military Thought], No. 6 (June 1967), trans. in Foreign Press Digest, No. 0112/68, July 11, 1968, p. 38.

60. Washington Post, 26 June 1967.

61. Raymond L. Garthoff, "SALT and the Soviet Military", Problems of Communism, Vol. 24, No. 1 (Jan.-Feb. 1975), p. 37.

62. Razvitie protivo-vozdushnoi orborony [Development of Anti-Air Defense], ed. by Marshal G.V. Zimin (Moscow, 1976), p. 191, cited in Michael J. Deane, "Soviet Military Doctrine and Defense Deployment Concepts: Implications for Soviet Ballistic Missile Defense," in Jacquelyn K. Davis et. al., The Soviet Union and Ballistic Missile Defense (Cambridge, MA: Institute for Foreign Policy Analysis, March 1980), p. 39.

63. Lieutenant General D.I. Shuvyrin, "A Reliable and Effective Defense," Voennye znaniia [Military Knowledge], No. 10 (Oct.) 1968, p. 17, cited in Michael J. Deane, Strategic Defense in Soviet Strategy (Coral Gables, Florida: Advanced International Studies Institute, 1980), pp. 39-40.

64. Razvitie protivo-vozdushnoi oborony, pp. 192 and 100, cited in Deane, "Soviet Military Doctrine and Defensive Deployment Concepts," p. 41.

65. U.S. Senate, Committee on Armed Serivces, Subcommittee on General Procurement, Soviet Defense Expenditures and Related Programs. Hearings, p. 84.

66. Ibid, p. 82.

67. Ibid, p. 128.

68. Ibid.

69. "Top Leaders of the Soviet Armed Forces," *Air Force Magazine*, Vol. 65, No. 3 (March 1982), p. 49.

70. *The Military Balance 1981/82*.

71. Central Intelligence Agency, National Foreign Assessment Center, *Soviet and U.S. Defense Activities, 1970-1979* (Washington, DC: Central Intelligence Agency, 1980), p. 5.

72. U.S. Senate, Committee on Armed Services, Subcommittee on General Procurement, *Soviet Defense Expenditures and Related Programs*. *Hearings*, pp. 122-123, 128.

73. "Brezhnev's Report to the Congress--I," *Current Digest of the Soviet Press*, XXXIII, 8 (March 25, 1981), p. 11.

74. Garthoff, "SALT and the Soviet Military," p. 22.

75. U.S. Senate, Committee on Armed Services, *Authorization for Military Procurement, Research and Development, Fiscal Year 1970, and Reserve Strength*. *Hearings*, Part 1, 91st Con., 1st Sess., 1969, pp. 8, 27.

76. Maj. Gen. I. Anureyev, "Determining the correlation of Forces in Terms of Nuclear Weapons," p. 44. Anureyev's probability coefficient for overcoming anti-air defense, for example, was 0.7 (or a 30% kill ratio), which is reasonable.

77. A.A. Sidorenko, *The Offensive*, trans. by United States Air Force (Washington, DC: GPO, 1972), p. 43.

78. Garthoff, "SALT and the Soviet Military," pp. 29-30.

79. Col. V. Bezzabotnov, "The U.S. Limited ABM System 'Sentinel'," *Voyennaya mysl'*, No. 5 (May 1978), trans. in Foreign Press Digest, No. 0013-69, Feb. 4, 1969, pp. 74-75.

80. U.S. Senate, Committee on Armed Services, *Authorization*, Part II, p. 1749.

81. Bezzabotnov, "The U.S. Limited ABM System 'Sentinel'," pp. 74-75; and William T. Lee, "Soviet Targeting Strategy and SALT," Air Force Magazine, Vol. 61, No. 9 (Sept. 1978), p. 120.

82. Andrei Gromyko, quoted in John Erickson, Soviet Military Power, Special Supplement to Strategic Review, Vol. 1, No. 1 (Spring 1973), p. 42.

83. See, for example, V. Kulish and S. Fedorenko, "Po povodu diskussii v SShA o strategicheskikh vooruzheniiakh," [On the U.S. Discussion of Strategic Arms] Mirovaia ekonomika i mezhdunarodnye otnosheniia [World Economics and International Relations], No. 3 (March) 1970, p. 48.

84. Raymond L. Garthoff, "Negotiating with the Russians: Some Lessons from SALT," International Security, I, 4 (Spring 1977), pp. 7, 11.

85. The PRC's facilities for missile storage, preparation, launch, and control are so primitive, and the U.S.S.R.'s intelligence and fast-reaction capabilities so sophisticated, that the Soviet Union might be able to react to China's activation measures with a take-out strike that would arrive on target before the Chinese missiles had left the ground. However, caution militates against eliminating one's backup system on the presumption of perfection. See C.G. Jacobsen, "SALT; MBFR: Soviet Perspectives on Security and Arms Negotiations," Center for International Affairs, Harvard University, Sept. 1973, p. 19.

86. Lt. Gen. G. Semenov and Maj. Gen. V. Prokhorov, "Scientific-Technical Progress and Some Questions of Strategy," Voyennaya mysl', No. 2 (Feb. 1969), trans. in Foreign Press Digest, No. 0060/69, June 18, 1969, p. 49.

87. ICBMs can be "pinned down" by striking command, control, and communications centers, and by delivering a powerful nuclear air burst over the ICBM site. Since in the latter two cases no hard targets are involved, low-accuracy weapons such as SLBMs are well suited to this mission.

88. For an interesting discussion of the role of interwar bargaining in Soviet strategy, see Gen.

S. Ivanov, "Soviet Military Doctrine and Strategy," *Voyennaya mysl'*, No. 5 (May 1969), trans. in Foreign Press Digest, No. 0116/69, Dec. 18, 1969, p. 49.

89. Garthoff, "Negotiating," p. 9.

90. Lewis A. Fran, "Soviet Power After SALT I: A Strategic-Coercive Capability?" *Strategic Review*, Vol. II, No. 2 (Spring 1974), pp. 57-58.

91. Bondarenko, *Sovremennaia nauka i razvitie voennogo dela* [Modern Science and the Development of Military Affairs], p. 132.

92. Sergei Konstantinov, "Neutron Bomb--Squaring the Circle," *New Times*, No. 35 (Aug.) 1977, p. 6.

93. Academician Markov, quoted in M. Chernousov, "The Voice of Soviet Public Opinion," *New Times*, No. 51 (Dec.) 1977, p. 6.

94. Lev Bezymensky, "N-Bomb Over Europe," *New Times*, No. 36 (Sept.) 1977, pp. 13-14.

95. S. Stashevskii and G. Stakh, "Kosmos dolzhen byt' mirnym," pp. 19-20.

96. Henry Kissinger, *White House Years* (Boston: Little, Brown, and Co., 1979), pp. 185-186.

5
Space-Based Lasers and the Evolution of Strategic Thought

Barry R. Schneider

If the technical problems are solved, space-based lasers (SBLs) might contribute to an effective defense against ballistic missile attacks. As Senator Malcolm Wallop has noted:

> We spend money, dollar after dollar and billions and billions, for weapons whose only consequence is to kill people. Now we have within our capability the possibility of developing weapons whose only real role in the world is to kill the things that kill people.[1]

Some advocates of space-based lasers believe they can become potentially decisive weapons that shift the advantage from nuclear offense back toward non-nuclear defense. Space-based lasers are potentially decisive strategic systems that are not weapons of mass destruction.

Advocates of space-based lasers believe that SBLs can help the United States to shift from strategic doctrines of assured destruction and countervailing strategy to strategic policies that emphasize non-nuclear strategic defense and assured survival.

Space-based lasers could revolutionize 21st Century warfare and thinking about national security problems to the same degree that nuclear weapons and ballistic missiles have changed warfare, force structures, and strategic thinking in the 20th Century.

After the first U.S. atomic explosions at Alamagordo, Hiroshima, and Nagasaki, it was clear that the explosive power of the new weapons were not "man-sized" any longer. "Little Boy" and "Fat Man" were fission bombs that were a thousand times more powerful than an equivalent weight of TNT bombs. Five years later the first thermonuclear bombs (H-bombs based on fusion) contained an explosive yield a thousand times more powerful than A-bombs. In half a decade the explosive power of weapons had multiplied a million fold. The past thirty-seven years of R&D improvements in yield-to-weight have resulted now in nuclear weapons that are 20 million times more powerful than conventional bombs.

This vast multiplication in destructive power meant that military defenses of urban areas and other "soft" targets had to become leakproof, for if even a small number of such weapons penetrated the results would be catastrophic. Entire cities or massed armies could be destroyed in minutes by the blast, firestorm, and radiation of a single weapon.

On the other hand, perfection may not be necessary. All BMD needs to accomplish is a shift in the overall fortunes of the attacker and defender in the direction of the latter. BMD is successful if it creates enough uncertainty about a positive outcome, that the potential attacker dares not attack.

It has been questionable, and still is, whether a sealproof defense against aircraft is possible, much less an effective defense against ballistic missiles ten times faster than supersonic bombers. Defense against ballistic missiles at first seemed impossible. Even after early anti-ballistic missile (ABM) systems in the 1960s showed that it was possible, under test-range conditions, for a "bullet to hit a bullet," U.S. defense officials concluded that it would be impossible for ballistic missile defenses to stop a sizable attack. Unable to defend against ballistic missiles that could tear holes through anti-bomber defenses, the United States decided not to try to defend against either ballistic missiles or bombers. The U.S. Defense Department essentially abandoned the idea and practice of strategic defense. Deterrence of war through the threat of nuclear retaliation became the mission of U.S. strategic forces.

EVOLUTION OF U.S. STRATEGIC DOCTRINE

The United States is pursuing laser technologies which, in the words of a recent GAO Report, "may revolutionize military strategy, tactics, and doctrine."[2] The Defense Advanced Research Projects Agency (DARPA) within the Office of the Secretary of Defense manages the research and development of three major SBL programs:

(1) Laser R&D (Alpha Program)

(2) Large Optics Demonstration Experiment (LODE)

(3) Target acquisition, tracking, laser aiming (Talon Gold)

These three programs are called the DARPA Space Triad. The success of these programs could alter the U.S.-U.S.S.R. strategic balance and would require a re-thinking of U.S. strategic doctrine.

As Dr. Maxwell Hunter has written:

> When lasers are placed in space so that every location on this planet is placed continuously in the target area of a laser battle system, then one has a right to expect truly fundamental changes. It raises the distinct possibility that the rapid delivery of nuclear explosives can be prevented by a weapon system which is itself not capable of mass destruction. Such a system would clearly give the nation that possesses it options in strategic posture and activity which are now denied everyone, including returning to the human beings in charge the time to permit adequate decision-making which was taken from them by the unholy synergism of nuclear weaponry and ballistic missiles.[3]

U.S. strategic doctrine since World War II has been partially determined by such factors as available military technologies, the size and quality of opposing strategic forces, and by domestic, bureaucratic, and international politics.

Changes in the technological state-of-the-art can present opportunities for doctrinal change and can also result in "doctrinal lags" when policy-makers are slow to see that their strategies have become anachronisms in the face of revolutionary technologies.

Nuclear weapons and the ability to deliver them from one continent to another via airpower changed U.S. strategic doctrine from emphasis on defense to a far greater reliance on deterring conflict through punitive threat.

During the 1945-1949 period of United States nuclear weapons monopoly, we also had a third policy option -- nuclear coercion -- that we declined to exercise. U.S. values did not include a strategic doctrine based on nuclear aggression. Nor did the size of the U.S. nuclear weapons arsenal encourage such strategic adventurism had we been so inclined. In 1945 the U.S. "had only two nuclear weapons in the stockpile; in 1946, nine; in 1947, thirteen; and in 1948 we had fifty."[4] Only after the first Soviet nuclear test in 1949 did the United States begin to build a much larger stockpile reaching a peak of several tens of thousands of nuclear weapons by the mid-1960s.[5]

THE FIFTIES AND SIXTIES: FROM MASSIVE RETALIATION TO FLEXIBLE RESPONSE

During the 1950s, while the United States still enjoyed a decisive nuclear superiority, declared U.S. policy was to threaten "massive retaliation" with nuclear strikes should the Soviet Union start a conflict with the United States or its allies. Once the Soviet strategic nuclear buildup had reached respectable dimensions in the late 1960s, U.S. strategic doctrine was modified to adjust to this new reality. Then the U.S. pursued "graduated deterrence" or "flexible response," whereby Soviet aggression would be met at its own level (conventional, theater nuclear, strategic nuclear) and defeated or stalemated at that level if possible. Escalation to the next level or beyond would be initiated if U.S. and allied forces were unable to contain enemy attacks at lower levels of conflict. This policy essentially linked the conventional defense forces of the NATO alliance with the U.S. strategic forces in such a way as to bring them into play as a last resort and as a deterrent to adversary initiation or escalation of conflict.

McNAMARA'S DEFINITIONS OF SUFFICIENCY

In the 1960s, the Kennedy-Johnson Administrations also announced a declaratory strategic deterrent doctrine called "assured destruction." Secretary of Defense Robert McNamara argued that the U.S. Government could deter any national Soviet Government from launching a military attack on the United States by maintaining a strategic nuclear force capable of riding out a surprise attack and then inflicting "unacceptable" damage to the aggressor. McNamara's operational definition of "unacceptable" damage in 1965 was damage equivalent to destruction of half the rival's industry and a quarter of its population.[6]

One of the advantages of the "assured destruction" doctrine was that the strategy could be implemented by a finite number of strategic nuclear weapons. DoD studies in the Sixties indicated that 400 one-megaton nuclear weapons exploded over Soviet targets would inflict 50 to 100 million deaths and would destroy at least 60 percent of Soviet industry. Beyond 400 such weapons, only marginal additional damage could be inflicted upon the U.S.S.R. through added nuclear strikes.[7] Since the Soviets had strategic nuclear force loadings that numbered in the hundreds in the 1960s and an estimated 1,300 by 1970,[9] both superpowers then possessed the capability to destroy the population and industrial centers of the other. This "mutual hostage" situation and doctrine is often referred to by its acronym MAD -- Mutual Assured Destruction. A MAD strategy was open to two fundamental critiques. First, critics argued that MAD was immoral. If implemented, it would kill millions of innocent civilians because they were unlucky enough to live in the U.S.S.R. and were ruled by leaders over whom they did not exercise much control, and who had initiated a nuclear strike. Second, a finite deterrence MAD strategy left no room for error. Once deterrence failed, the U.S. President was left only two choices -- surrender, or a full countervalue retaliation that would also doom U.S. cities once the war escalated to such levels.

U.S. strategic nuclear policy has always operated on four levels: employment policy, acquisition policy, declaratory policy, and deployment policy.[10] During the 1960s and early 1970s U.S. declaratory policy emphasized "assured destruction" as a method of deterring Soviet adventurism. Yet in the Single Integrated Operating Plan (SIOP), the U.S. war plan, "counterforce targeting has always been part of U.S. strategic planning (as exemplified in the SIOP since 1961),"[11] and as the numbers of U.S. strategic nuclear weapons grew in numbers and accuracy the growing majority of SIOP targets were Soviet military forces and installations.

THE SCHLESINGER DOCTRINE: LIMITED NUCLEAR OPTIONS

In 1974, the Administration advanced a "new" modification of the doctrine of assured destruction. Secretary James Schlesinger's publicly stressed the need to be able to fight limited nuclear engagements against Soviet military targets. Such "warfighting" capability would provide greater numbers of options short of all-out mutual annihilation should deterrence fail, and, it was argued, such a capability would be more credible to Soviet leaders and thus would improve U.S. ability to deter Soviet aggression.

The "MAD" strategic policy that had been articulated by Secretary McNamara in the 1960s recognized the idea of "sufficiency." A certain finite number of nuclear weapons could destroy the designated targets. The Schlesinger doctrine emphasized limited counterforce capabilities as important to deterrence and also added an additional guideline for restructuring U.S. strategic forces, "essential equivalence." Secretary Schlesinger stated that an additional essential element in the U.S. effort to deter nuclear war is that no perceived asymmetries in U.S. and Soviet capabilities should exist. This standard of "essential equivalence" is the intellectual linchpin to the present countervailing strategy embodied in PD-59. The U.S. and Soviet forces more and more are becoming the mirror image of each other, with an emphasis on counterforce warfighting capability as embodied in highly-accurate ballistic missile forces.*

During Secretary Schlesinger's tenure in office, President Gerald Ford also approved another refinement of U.S. targeting and deterrence policy. National Security Decision Memorandum 242, signed on January 17, 1974, reportedly ordered U.S. forces to be targeted in such a fashion that after any nuclear weapons exchange, the United States was to be able to exceed the Soviet Union in post-war recovery.

In action policy, the Schlesinger emphasis on limited nuclear options changed little in the early 1970s. The addition of the Command Data Buffer System permitted the rapid retargeting of missiles, but the SIOP itself was not dramatically changed. The main changes were in the increased emphasis on selectivity and flexibility and the provision of groups of just a few targets for Presidential choice, such as Soviet oil production capability. In the late 1970s the new, more accurate _Minuteman_ III reentry vehicles, with double their previous yield, made precision targeting a possibility.

For example, a recent OTA study on _The Effects of Nuclear War_ calculated that an attack on 24 Soviet

---

* Convergence theory has worked in reverse of what was expected. Rather than persuading Soviet strategists of the error of their ways, the U.S. has adopted a strategic doctrine, and now is committed to a force structure strikingly like that of the Soviets in its ability to strike hardened targets in a time-urgent manner.

refineries and 34 Soviet petroleum storage sites by the weapons of 7 Poseidon SLBMs and 3 Minuteman III ICBMs would destroy an estimated 73 percent of Soviet refining capability and 16 percent of Soviet oil storage capacity.[12] Such an attack might kill as "few" as 836,000 Soviet people and possibly as many as 1,458,000.[13]

The effect of a full counterforce exchange between the U.S. and Soviet Union has also been calculated. Executive branch countersilo studies indicate that "between 2 million and 20 million Americans would die within the first 30 days after an attack on U.S. ICBM silos."[14] Soviet fatalities have been calculated for a somewhat different scenario -- a U.S. retaliatory counterforce strike after a Soviet first-strike attack. DoD studies indicate Soviet fatalities in this case to be between 1 to 4 percent of the population (2.5 million to 10 million dead). Another study done by the U.S. Arms Control and Disarmament Agency estimated Soviet fatalities after a U.S. counterforce strike at 3.7 to 13.5 million persons dead.[15]

While these figures are forbidding, they are considerably less so than the estimated casualties from a full "assured destruction" (MAD) attack, an exchange where population centers, industrial complexes, and other non-military targets are destroyed. As many as 165 million Americans and 100 million Soviet citizens are estimated to die if each side struck a full set of adversary targets with their respective strategic nuclear forces.[16]

The "Schlesinger Doctrine" began a strategic debate that has never abated. Critics of the "warfighting preparation" school of deterrence fear that it encourages field commanders to think in terms of using nuclear weapons and might encourage policymakers to give nuclear release authority earlier than necessary when faced with conventional or insurgency conflicts where victory was not in sight. Once begun on the path up the escalation ladder, once the nuclear threshold has been breached, most strategic analysts fear escalation will continue until both warring countries lie in ruins with a death toll beyond historical experience.

HAROLD BROWN'S COUNTERVAILING STRATEGY: PD-59

This debate over "warfighting" versus "MAD" nuclear strategic policies was given fresh impetus in August 1980 when Secretary of Defense Harold Brown announced a "new" American targeting policy in a speech at the U.S. Naval Academy. Brown unveiled a new U.S. strategy -- the countervailing strategy -- which was based on President Jimmy Carter's approval of Presidential Decision Memorandum 59 (PD-59) in the Summer of 1980.

"The overriding objective of our strategic forces is to deter war," Brown stated, but added that through added U.S. counterforce weapons and new emphasis on counterforce targeting he wanted to "ensure that the Soviet leadership knows that if they chose some intermediate level of aggression, we could, by selective, large (but still less than maximum) nuclear attacks, exact an unacceptably high price."[17]

PD-59 focuses on the need to deter the Soviet leadership from either limited or all-out nuclear attacks by maintaining U.S. nuclear forces and C^3I assets capable of denying the Soviets achievement of their objectives at any level of conflict, or by inflicting costs upon them exceeding any of their anticipated gains. PD-59 thus requires U.S. forces capable of fighting effectively at each rung on the escalation ladder and forces that are able to endure repeated exchanges over an extended period of time.

PD-59 is based on the view that the most effective way to deter Soviet nuclear aggression is by maintaining escalation dominance, and by targeting those things which their leadership values most. The countervailing strategy would put in jeopardy the Politburo's political and military control, their conventional and military forces, Soviet defense industries, and the lives of Soviet leaders.

Deterrence is based on the idea that the defender could inflict an unacceptable level of damage on the attacker. But no one quite knows what Soviet leaders think is "unacceptable" damage. Proponents of the "countervailing" strategy believe Soviet leaders would be most deterred by the fear of defeat of their war machine, their loss of control at home, and by the threat of U.S. "decapitation" strikes against Andropov, Gromyko, Ustinov, and their Politburo colleagues in their bunkers. Thus, even if they were willing to sacrifice large numbers of their own people in war, they would still have a political and personal stake in seeing that no conflict occurred.

As former Secretary Harold Brown has stated:

> The biggest difference, I would say, that PD-59 introduces is a specific recognition that our strategy has to be aimed at what the Soviets think is important to them, not

just what we think would be important to us in their view.[18]

The countervailing strategy also is thought to put several additional rungs on the escalation ladder, options short of all-out nuclear war should deterrence fail.

Finally PD-59 and the new strategy are believed by some to enhance deterrence "across a range of possible threats for which large-scale economic destruction would be an unbelievable retaliatory option."[19]

Critics of PD-59 have voiced a variety of concerns about the new strategic doctrine, namely, that:

-- the focus on nuclear warfighting capability will make nuclear war more thinkable, therefore more likely. A diplomacy of nuclear brinksmanship may develop.

-- the doctrine "requires" a host of new U.S. counterforce weapons (MX, Trident II, ALCM, GLCM, SLCM) which will accelerate the U.S.-U.S.S.R. strategic arms race, spur Soviet counter-responses, and undermine crisis stability.

-- the focus on "decapitation attacks" versus Soviet leaders in wartime could eliminate the slim possibility of halting a U.S.-Soviet nuclear exchange short of mutual annihilation.

-- the tension between the desire to acquire new counterforce weapons and the desire for deep cuts in U.S. and U.S.S.R. strategic forces and intermediate nuclear forces might be resolved in favor of arms building rather than arms reduction.

-- "the more obvious 'warfighting' oriented declaratory policy of PD-59 cannot solve the self-deterrence dilemma unless it includes a commitment to effective damage-limitation and damage-denial."[20]

-- it still fails to address the illogic of threatening to initiate a process leading to national suicide on behalf of distant allies.

STRATEGIC DOCTRINE IN THE REAGAN ADMINISTRATION

The Reagan Administration has adopted and

extended the strategic doctrine adopted by President Jimmy Carter in his last year in office. President Reagan and Defense Secretary Weinberger have indicated that the countervailing strategy, Presidential Directive 59 is alive and well on their watch.

Several other Carter directives, PD-41 and PD-48, have also been adopted by Ronald Reagan as his policy. Presidential Directive 41 reportedly was concerned with the need to improve United States command, control, communication, and intelligence during wartime. Presidential Directive 48 reportedly ordered a new search for cost-effective means of providing active and passive defenses.

The Reagan Administration has also accepted the premise, in their first five-year defense plan, that nuclear conflict could be protracted and have begun planning against that contingency. The Weinberger Pentagon also sees a need for enduring nuclear and C^3I capabilities to deter a wide spectrum of threats.

Secretary of Defense Casper Weinberger, in his FY 1983 Annual Report to the Congress, stated that U.S. strategic nuclear forces are structured to perform multiple roles:

-- to deter nuclear attack on the United States and its allies.

-- to help deter major conventional attack against U.S. forces and our allies, especially in NATO.

-- to impose termination of a major war on terms favorable to the United States and our allies, even if nuclear weapons have been used.

-- to deter escalation in the level of hostilities (once begun).

-- to negate possible Soviet nuclear blackmail against the United States or our allies.[21]

Thus, the U.S. Government officially accepts the responsibility of deterring a very wide spectrum of threat through the deployment and counterthreat of U.S. strategic nuclear programs.

Other elements of U.S. strategic doctrine that have gotten additional emphasis and funding during the Reagan Administration[22] include:

-- providing strategic force endurance in the contingency of protracted war.

-- improving survivability and endurance of C^3I assets.

-- limiting damage to the United States if war occurs.

-- improving the ability to destroy military and war-support assets of the Soviet Union.

-- decreasing strategic force vulnerability to a surprise attack.

-- developing anti-satellite weapons to target Soviet space assets.

-- developing weapons to defend U.S. satellites and other U.S. space assets.

According to a story leaked to the New York Times in May 1982:

> Defense Department policy-makers, in a new five year defense plan, have accepted the premise that nuclear conflict with the Soviet Union could be protracted and have drawn up their first strategy for fighting such a war.[23]

The Reagan Administration has shown an interest in strengthening U.S. strategic forces and command, control and communications (C^3) so that they would be capable of sustained efforts over days, weeks, and perhaps as long as six months in a protracted nuclear-conventional war. Soviet military doctrine emphasizes the need for protracted operating ability and key Reagan Administration officials believe that deterrence would be better served if the U.S. was capable of countering the Soviet Union in such a conflict.

In the "Fiscal Year 1984-1988 Defense Guidance" military planners in the Reagan Administration reportedly state that, "nuclear war strategy would be based on what is known as decapitation, meaning strikes at Soviet political and military leadership and communications lines".[24]

According to reports,

> In developing a strategy for fighting a protracted nuclear war, Mr. Weinberger's policy

> planners went beyond President Carter's Presidential Directive 59, which focused on specific military and political targets.
>
> The new nuclear strategy calls on American forces to be able to "render ineffective the total Soviet (and Soviet-allied) military and political structure." But it goes on to require the assured destruction of "nuclear and conventional military forces and industry critical to military power. Those forces must be able to maintain through a protracted conflict period and afterward, the capability to inflict very high levels of damage on Soviet industry."
>
> The nuclear strategy emphasizes communications, so the President and his senior military advisers could control a nuclear exchange and not be limited to one all-out response to a Soviet attack.[25]

New U.S. plans call for numerous means for modernizing and protecting U.S. satellites and other C^3I assets including continued R&D funding of space-based satellite defenses.

The defense guidance also reportedly accelerated the development of offensive forces such as the Trident II D-5 missile that promises to be effective against Soviet missile silos and command bunkers. On the defensive side, Reagan defense planners wish to speed up research and development of U.S. ballistic missile defense programs. They are weighing whether to seek revisions of the ABM Treaty if deployment of the MX missile in a new basing mode required BMD to make it adequately survivable.

According to the *New York Times*, the defense guidance said that the Soviet Union "will have acquired, a decade hence, the capabilities to keep our land-based surface and naval forces under near constant surveillance, locate units and facilities precisely throughout the length and breadth of the area of operations, and engage those units and facilities in near real-time."[26] The U.S. would continue to develop its anti-satellite weapons in order to have the ability to combat this Soviet reconnaissance activity.

Several of the Reagan policy doctrines have had their initiation in the previous Administration. For example, the Carter Administration promulgated in 1980 not only the countervailing strategy found in PD-59, but also reportedly initiated PD-41 which emphasized the necessity for an improved U.S. civil defense program, PD-48 which reportedly renewed the U.S. search for improved cost effective active (BMD) and passive (civil) defenses, PD-53 which reportedly ordered improvements in U.S. C^3I between National Command Authorities and strategic forces, and PD-58 which reportedly ordered improvements in plans and procedures to provide continuity in government during a nuclear war. The Reagan Adminstration has embraced these improvements in force survivability and connectivity which could improve U.S. capabilities to limit damage and ensure survival and leadership during a counterforce nuclear exchange.

However, the Reagan Administration has not yet gone beyond marginal steps from a pure strategy of deterrence to what some call "classic strategy" with its emphasis on "real defense" of the U.S. homeland. Actual strategic defense of the continental and offshore United States territory would require a workable and very comprehensive ballistic missile defense. Space-based lasers might be a key to such BMD possibilities.

SBLs: PRESENT AND FUTURE IMPLICATIONS

Space-based lasers may be used as future (1) ballistic missile defenses, (2) anti-satellite weapons, (3) as satellite defenses and (4) as defenses against bombers. Effectiveness in any of the three roles could make officials change U.S. strategic doctrine to take advantage of SBL possibilities.

SBL-BMD

The United States, according to the 1972 ABM Treaty as modified by the 1974 ABM Protocol, can deploy 100 ABM interceptor missiles at one ICBM site. One initial site choice was at the *Minuteman* base in Grand Forks, North Dakota, and we briefly deployed *Spartan* ABMs there in 1976. After 10 months of operation, U.S. policy changed and the ABM launchers were dismantled. The Perimeter Acquisition Radar was turned over to NORAD for integration into the U.S. early warning radar network. Since 1976 the U.S. has deployed no ballistic missile defense system.

The *Safeguard* ABM site was dismantled because it was considered ineffective against MIRVed Soviet ICBMs and not worth the costs of maintaining such a marginal system.

Since 1972 the United States has twice reviewed the ABM Treaty and decided to continue its provisions in force. According to news reports based on leaks of the classified defense guidance for the years 1984-1988, the United States Government is committed to accelerating research and development work on ballistic missile defense systems.

The U.S. Government might also seek to revise the ABM Treaty if the MX missile were deployed so that it required an ABM force to help it survive a Soviet first-strike.

The United States is exploring a number of BMD technology options. The most mature and promising is the Low-Altitude Defense System (LoADS) which could carry either nuclear or non-nuclear kill (NNK) interceptors. This endoatmospheric BMD could be deployed first, perhaps as early as 1990. In the 1990s, the U.S. might be able to deploy an effective exoatmospheric ballistic missile defense system. This BMD force would be designed to intercept attacking missile or RVs in space before they began their descent through the earth's atmosphere toward their targets below in the continental United States.

To this two-layered U.S. ballistic missile defense system, based on missile and explosive technologies, might be added a space-based laser third layer of defense. As a third and outer layer of defense, space-based laser BMD need not work with great effectiveness to be of value. The combined system could work synergistically, and create far more problems for adversaries engaged in concocting effective countermeasures.

Space-based laser ABMs would require extensive modification to the ABM Treaty, or would require its termination. Because of the absence of an effective near-term ABM technology, the Reagan Administration has been unwilling to lose the benefits of the ABM Treaty.

As the number and accuracy of offensive missile reentry vehicles have increased, fixed-site targets like ICBM silo-launchers have become progressively more vulnerable. Also less secure are launch control centers, leadership shelters, Emergency Rocket Communications Systems (ERCS) and most fixed U.S. C^3I assets. Anti-ballistic missiles and/or space-based laser BMDs may provide a shield for some of these assets.

Space-based lasers might, by the late 1990s, provide the outer layer in a three-tiered ballistic missile defense. An SBL-BMD with its speed-of-light capability could engage the Soviet missile force as soon as it had climbed from launch pads and escaped the earth's atmosphere. SBL kills of Soviet missiles prior to release of MIRVs could simplify the tasks of LoADs by thinning out the attack considerably. One SBL-BMD "kill" of an SS-18 would destroy 10 MIRVs and might preserve 5-10 U.S. ICBM silos. SBLs could provide a forward defense of ICBMs.

The second and third components of a U.S. BMD system could consist of nuclear or non-nuclear missile-interceptors. Exoatmospheric interceptors could help space-based lasers with the task of thinning the Soviet attack force before it released its multiple independently targetable reentry vehicles. This one-two punch in space could greatly simplify the job of U.S. low-altitude interceptors by reducing the size of the attack force that leaked through the outer defenses.

A LoADs BMD that stopped only 10 percent of the attacker's reentry vehicles might be considered inadequate in a stand-alone mode. But LoADs, complemented by SBL-BMDs and exoatmospheric interceptors that destroyed much of the other 90 percent of an attack force, could be invaluable. The multiplicity of U.S. BMD systems might also make it more difficult for the Soviets to design forces and tactics as countermeasures. What could work against one, might not against the others.

The better that BMD systems can thin out an attack, the greater the survivability potential certain key C^3I assets such as launch control centers for _Minuteman_ complexes and Emergency Rocket Communications Systems (ERCS) that reportedly are fired East and West from _Minutemen_ II silos in the event of attack. ERCS signals carry the emergency action message (EAM) of national command authorities to U.S. bombers that are traveling toward their turnaround points, to U.S. ICBMs and to TACAMO (Take Charge and Move Out) aircraft vital for relaying messages to U.S. fleet ballistic missile submarines. BMD might also provide a degree of safety for preferentially defended U.S. leadership shelters. Although the Reagan Administration has yet to view BMD in terms other than a defense of land-based ICBMs (MX and _Minuteman_), a proven hard point BMD technology might be a way station on the road to a later area ballistic missile defense. A credible defense of U.S. cities might become a national strategic objective and an eventual reality, but if only an extremely effective SBL-BMD technology can be developed.

The present U.S. BMD thinking in official circles remains focused on deterrence, and on point defense of the retaliatory forces. Nevertheless, the trend in both plans and programs, if continued long enough, could provide the means and thinking for "shifting the strategic paradigm from mutual assured destruction to damage-limitation. Space-based lasers might, in combination with other BMD technology, make this feasible. This shift to strategic defense would be congruent with recent DoD decisions strengthening civil defense and continental U.S. air defense and the renewed attention in the Pentagon to the strategic advantages of a BMD deployment beyond the defense of ICBM forces."[27]

Space-based laser battle stations would not need to operate perfectly to make potential aggressors think twice about attacking the United States. The preservation of a single MX ICBM from destruction could lead to the destruction of three or four major Soviet cities or elimination of key industrial resources such as Soviet oil refining and storage capability.

SBL-BMDs might provide additional time for U.S. decision-makers to consider responses to attacks. Instead of 15 minutes from warning to decision, the U.S. President might have hours or even days to respond. SBL-BMDs, if they worked with high efficiency, could buy time for cooler decisions and more crisis bargaining.

An effective SBL-BMD could allow the United States to avoid going to a "launch-on-warning" or "launch-under-attack" posture before hostilities had occurred. If SBL-BMD protection of U.S. ICBMs and airbases helped avoid a "use or lose" situation, they could contribute to avoiding accidental nuclear war, and also to more considered U.S. military responses to attacks. This might limit the escalation of conflicts. Should an adversary attack U.S. space-based laser stations without also proceeding to attack other U.S. forces, the U.S. might then respond by placing its ICBMs in a launch-on-warning status and making this known to the enemy. This might be accompanied by a tit-for-tat destruction of a prized enemy asset.

It should be noted that a highly effective space-based laser weapon may be able to defend itself first and then carry on in its BMD role.

Effective SBL-BMD systems could improve deterrence by protecting U.S. retaliatory forces, discouraging potential aggressors from launching the first attack in a war they would be unlikely to win. Should deterrence fail, effective SBL-BMDs might reduce the damage and casualty rates and help preserve ground-based C^3I links and enable the U.S. to continue a more extended conflict. Thus, SBL-BMD R&D

would be consistent with the strategic doctrines of the Reagan and Carter Administrations.

<u>SBL-ASATs</u>

Space-based lasers could also serve as anti-satellite (ASATs) weapons or as defenses for satellites (DSATs). Once a SBL battle station got warning of a Soviet missile attack, its beams could blind Soviet satellites useful to their military. For example, laser beams could be used to destroy Soviet <u>Molniya</u> and <u>Cosmos</u> communications satellites, severing lines of communication between Moscow and its world-wide military units. SBLs could destroy, in a few minutes, Soviet navigation satellites and ocean surveillance satellites. SBL battle stations might also be able to blind Soviet reconnaissance and early warning satellites. After the first U.S. SBL salvos, the Soviet leadership could be in the dark about the outcome of their attack. A unilateral U.S. advantage in space-based laser weapons would have the potential to blind the satellite eyes of the Kremlin. Such an SBL-ASAT capability would improve U.S. abilities to deter war, to establish escalation dominance during wartime, and to prolong U.S. capabilities for waging war while reducing the Soviet threat to U.S. space-based C^3I linkages.

It is unlikely that either the United States or the Soviet Union would permit the other to keep a monopoly on ASAT weapons. If either side develops a very effective ASAT, both will attempt to do so; indeed the Soviet Union now has an operational ASAT system and the U.S. is developing a conventional ASAT system for mid-1980 deployment.

<u>SBL-DSATs</u>

Similarly, a U.S. space-based laser weapons force would need to provide for its own defense against attack, while providing a speed-of-light defense for U.S. satellites being attacked by rocket ASATs from the Soviet Union. SBL-DSATS might be used to protect the following kinds of U.S. space assets:

<u>U.S. SATELLITE ASSETS</u>[28]

<u>Reconnaissance</u>: Big Bird Photo-Reconnaissance Satellites
KH-11 Photo-Reconnaissance Satellites
Electronic Reconnaissance Satellites

Communications:	Defense Satellite Communications Satellites
	Fleet Satellite Communications Satellites
	Leased Satellite System
	Air Force Satellite Communication System
	MILSTAR Satellites (1980s)
	Satellite Data System Satellites
Meteorology:	TIRUS Satellite
	GOES Satellites
	Nimbus Satellites
	NOAA Satellites
	Advanced Meteorological Satellites
	Geostationary Operational Environmental Satellites
Navigation:	Transit Satellites (U.S. Navy)
	NAVSTAR GPS Satellites (All Services)
Nuclear Explosion Detection:	Vela Satellites
	NAVSTAR GPS Satellites
Ocean Surveillance:	ELINT Ocean Reconnaissance Satellites
	Navy Ocean Surveillance Satellites
	NASA/SeaSat Satellites
Early Warning:	Teal Ruby Satellites (1983)
	Vela Satellites
	RH Satellites
	T-3C Satellites
	IMEWS Satellites

SBL-DSATs might help preserve the C^3I satellite links between authorities and U.S. forces. They could help to preserve satellites whose assistance to U.S. forces are "force multipliers" in the field. Thus, SBL-DSATs could make possible continued U.S. war-fighting capability longer into the conflict and might the space-based assets needed to acquire enemy targets and assess the effects of exchanges. This preservation of battle management capabilities using DSATs can serve to limit damage done in wartime by enemy strategic forces. SBL-DSAT protection of U.S. C^3I satellites would be consistent with strategic doctrine of the Reagan Administration which emphasizes the need to improve U.S. command, control, and communications connectivity and continuity. Of course if all ground-based C^3I nodes and downlinks from U.S. C^3I satellites were not effectively protected, the United States might still lack the communications needed by a President and his senior military authorities to coordinate a counterforce retaliatory response rather than a single spasm response to a Soviet nuclear strike. Similarly a U.S. ASAT advantage could help deny the same C^3I to Soviet leaders and would be consistent with the defense guidance for 1984-1988, as well as with PD-59 which calls for targeting Soviet political and military leadership and lines of communication.

Unfortunately, a system of Soviet spaced-based lasers might endanger the constellation of U.S. satellites unless Soviet SBLs were disabled very early in a U.S.-U.S.S.R. conflict. Preemption against the rival SBL battle stations could be a first order of business for any nation engaged in war and could be a great temptation in a crisis that appeared to be on the brink of war.

SBL Bomber Defenses

Space-based lasers might also be used to defend against future enemy bomber attacks in two ways. First, SBL weapons could be used as the kill mechanism that directly destroys enemy bombers. Second, space-based lasers might simply "lock on" to the enemy bombers and provide the designator beam that U.S. air-to-air missiles follow to target.

Future SBL-bomber defenses would likely be most effective, if at all, at the higher altitude where the earth's atmosphere would least intefere with coherent propagation of the laser beam. Bombers perceiving a laser attack might counter by flying at lower altitudes, accepting decreased range penalties for the benefits of increased protection.

Space-based Lasers and C^3 Protection

If used to help defend U.S. ICBMs long enough to permit effective retaliation against the Soviet Union in the event of a Soviet first-strike, laser weapons would add to stability. If SBL-DSATs added protection to U.S. satellites this could help protect the space-based C^3I assets of the United States needed to guarantee effective military operations in wartime. Soviet knowledge of this should improve the U.S. ability to deter them from attack.

Even a marginally effective U.S. SBL-BMD system might be useful if it was used to protect key U.S.

assets. SBLs might, for example, be used not only to protect U.S. ICBMs housed in their silos but also to blunt attacks against U.S. national command authorities at the White House, the Pentagon, SAC Headquarters and NORAD. SBLs might thin out attacks that threatened the Emergency Rocket Communications System (ERCS) and other vital C^3I links between national command authorities and strategic forces.

Advantages of a "Thin" Defense

A SBL-BMD shield would have advantages over a nuclear-armed ABM force. It would not require a real-time Presidential decision to release nuclear armed interceptors since no nuclear weapon or other mass destruction device is involved.

A marginal SBL-BMD shield also could help protect the United States from a "light" ballistic missile attack from a third party, helping the U.S. avoid a catalytic nuclear war and permitting time to assess from whence the strike came. Also, SBL weapons could be a form of insurance against the unauthorized launch of a missile or two by either U.S. or Soviet military officers acting on their own, by intercepting them before they triggered a nuclear war.

SBLs and Adversary Reactions

The chances are that if one superpower develops effective space-based laser weapons, the other will not be far behind. A defense-dominated world will probably be two-sided. As one BMD expert has written, "In real life, either both nations will have significant BMD deployments or neither nation will have them."[28]

An effective SBL-BMD system would be a high priority target in either the U.S. or Soviet warplans. Whoever destroyed the rival SBL-BMD first, would stand to gain heavily in the ensuing military battle. Decision-makers under pressure in crises may view SBL weapons, much in the same fashion they might view nuclear weapons about to be overrun, and face the same "use or lose" dilemma.

Preemption against U.S. space-based laser battle stations would probably be the first order of business should the Soviet Union decide to war against the United States. First, it is Soviet strategic doctrine to hit first with its offensive might to weaken the adversary's offensive and defensive capabilities. Second, this is only logical if the U.S. space-based laser weapons were capable of neutralizing much of their attacking forces. Removal of the first line of U.S. defenses would open more targets to attack.

It is unlikely that deployment of space-based battle stations would make them solitary targets to be put out of commission in a rough undeclared war in space that did not entail land, sea, and air combat as well.

Nevertheless should either superpower place laser-BMD battle stations in space, shielding and defense of the battle stations would be a top priority.

U.S. strategic doctrine might then have to be modified to emphasize the need to protect the first line of strategic defense, the SBL-BMD. Just as defense of ICBMs is important today and SBL-BMD could be an antidote to ICBM vulnerability, it eventually may be necessary to protect the ICBM protector. Just as retaliatory forces are today preserved by deception, mobility, hardening, dispersion, proliferation, and active defenses, so too might such measures be required to preserve SBL-BMDs. Not unlike aircraft carriers designed for power projection but using increasing portions of available resources for self-defense, SBL-BMDs may need their laser capability or their own space convoys to protect them in their space ocean.

SUMMARY AND CONCLUSIONS

The idea of strategic defense of the United States faded as the Soviet Union developed the new technologies of nuclear explosives and ballistic missiles. Strategic defense was replaced by the doctrine of strategic deterrence.

Ironically, "the dangerousness of war has reduced the danger of war,"[30] since nuclear war has appeared too awful to risk for less than absolutely vital interests. This does not mean that conventional wars are not possible, but it does mean that nuclear-armed rivals must deal with each other very carefully indeed.

The synergism of long-range airpower and nuclear physics has rendered the nation-states of the world much less defensible. Instead of being able to meet invaders at the borders on battlegrounds at a distance from the core of the country, today every U.S. citizen (and every Soviet citizen) is "at the front." We are each thirty minutes away from an annihilating attack that we are presently powerless to stop.

New ballistic missile technologies including rocket-boosted interceptors as well as space-based laser weapons offer a potential solution to this security dilemma. If laser technology R&D overcomes some extremely difficult problems in the next two decades, the United States may, for the first time since the Soviets acquired thermonuclear weapons, be able to defend against an actual attack. This would, at a minimum, help us protect our retaliatory forces and C^3 that constitute our deterrent to war.

Space-based laser weaponry may be the vehicle whereby the advantage is returned to defensive forces after decades of offensive predominance. Nuclear deterrence has helped to keep a lid on war, but how long can our luck continue to hold? Deterrence need fail only once for an unparalleled world catastrophe to occur. Strategic thinking must find a new route to security. One path is deterrence of war by maintaining a retaliatory capability feared by an adversary. Another path is nuclear disarmament and detente. Still another is acquiring a strategic defense that works with high efficiency. The three approaches can be complementary and reinforcing. The last two are destinations reachable only after hard journeys.

The first effective U.S. ballistic missile defenses, anti-satellite weapons, and satellite defenses are likely to be dependent on more mature and cost-effective technologies than will be available via space-based lasers. Laser weapons will initially serve to supplement other existing forces in these roles. But even a marginally effective laser weapon operating in space might be enough to shift the advantage from the attacker to the defender when combined with other ballistic missile defenses. Laser speed-of-light capabilities would allow them to engage enemy ballistic missiles sooner and longer than rocket-boosted interceptors. SBL-BMD kills of enemy missiles early in their flight trajectory, before MIRV separation, could thin out the attack so that low-altitude BMD interceptors could handle the residue of the attack force. The missile/bomber threat these SBL weapons will face in the 21st Century could be reduced considerably if the START and INF negotiations succeed in reducing nuclear arsenals to "more defendable" levels. If nuclear arsenals were reduced by about one-third as President Reagan suggested in the U.S. START proposal, then defenses would have an easier task to perform.

An important mission for space-based lasers and other anti-ballistic missiles forces might be preserving key elements in the U.S. national command structure and C^3I network to avoid the possibilities of Soviet decapitation attacks that might paralyze U.S.

retaliation. SBL-BMDs and other ballistic missile defenses might be used to buy protection for command, control, and communications aircraft such as Looking Glass, the National Emergency Airborne Command Post (NEACP), the Post-Attack Command and Control Systems (PACCS), and the Take Charge and Move Out (TACAMO) systems. Space-based laser battle stations and other ballistic missile defenses might also be employed to defend hardened bunkers housing U.S. national command authorities and the Minuteman silos housing the Emergency Rocket Communications System (ERCS) missiles. Defense of these C^3I assets would improve the U.S. ability to retaliate and prosecute a war. Soviet knowledge of that fact should improve the U.S. deterrent posture.

While space-based laser weapons remain in their R&D infancy they are unlikely to exert much influence on U.S. or Soviet force structures or strategic doctrines. Eventually, however, space-based laser weapons, together with other advances in strategic defensive techology, might allow the U.S. to alter its strategic doctrine from one that would preserve the peace simply through the threat of nuclear retaliation to a new policy that emphasizes defense and deterrence in equal measures.

NOTES

1. "Senate Directs Air Force to Formulate Laser Plan", Aviation Week and Space Technology, May 25, 1982, Vol. 114, No. 21, p. 53.

2. Comptroller General of the United States, "DoD's Space-Based Laser Program -- Potential, Progress, and Problems, (Wash. DC: GAO, Feb. 26, 1982), C-MASAD-82-10. (UNCLASSIFIED)

3. Maxwell W. Hunter, II, "Strategic Dynamics and Space-Laser Weaponry," Oct. 31, 1977, Lockheed Missiles & Space Company, Inc., Sunnyvale, California.

4. U.S. Department of Defense, News Release, "The United States Nuclear Weapon Stockpile," June 1, 1982. See also: Congressional Record, Proceedings and Debates of the 97th Con., 2nd Sess., Vol. 128, No. 106, Part II, Wash. DC, Thursday, Aug. 5, 1982, p. H5334.

5. Ibid.

6. William Arkin, Thomas Cochran, and Milton Hoenig, "U.S. Nuclear Stockpile: Materials Production and the New Weapons Requirements," Arms Control Today, April 1982.

7. See testimony, Secretary of Defense Robert S. McNamara, Defense Subcommittee of Appropriations Committee, U.S. Senate, Feb. 24, 1965, Department of Defense Appropriations, 1966, (Wash. DC: GPO, 1965), Part I, pp. 44-45. McNamara's criteria for assured destruction changed over the years in the Pentagon.

8. Alain Enthoven and Wayne Smith, How Much Is Enough? (New York: Harper and Row, 1974), pp. 207-210.

9. The United States had 4,600 Strategic Nuclear weapons in 1970, the Soviets 1,300 according to various Annual Reports of the Secretary of Defense.

10. As Lynn Davis, a former staff member of the U.S. National Security Council and former Asst. Sec. of Defense has explained, Employment policy describes the targets and how the United States

plans to use the nuclear weapons which it possesses today. Acquisition policy establishes criteria for developing and procuring nuclear weapons for the future. Declaratory policy gives guidance to American officials on what they say publicly about the employment and acquisition policies. Deployment policy designates where nuclear weapons are to be stationed." See: "Limited Nuclear Options, Deterrence and the New American Doctrine," Adelphi Papers, No. 121 (London: International Institute for Strategic Studies, Winter 1976/76) p. 1.

11. Desmond Ball, "Déjà Vu: The Return to Counterforce in the Nixon Administration," Dec. 1974, California Seminar on Arms Control and Foreign Policy, p. 45.

12. Office of Technology Assessment, The Effects of Nuclear War "Case 2: A U.S. Attack on Soviet Oil Refineries" (Wash. DC: GPO, May 1979), p. 76.

13. Ibid.

14. Ibid., p. 84. The variation is due to uncertainties about the scenario and the type of Soviet attack. Key uncertainties center around height of weapons burst, weapon's design, wind, rain, terrain, and distance of the population from the explosions.

15. Ibid., p. 91.

16. Ibid., See p. 100 for Soviet casualty figure and p. 95 for U.S. casualty estimates. These are estimates of deaths in the first 30 days of a "spasm" nuclear war.

17. Harold Brown, speech at the U.S. Naval Academy, Aug. 20, 1980. For a summary of PD-59 and a critique of the doctrine see: Paul C. Warnke and Barry R. Schneider, "A Nuclear War Must Never Be Fought," Across the Board, March 1981, pp. 4-10.

18. Harold Brown, testimony before Committee on Foreign Relations, U.S. Senate. See: Nuclear War Strategy, Hearings, 96th Congress, 2nd Sess. (Wash. DC: GPO, Feb. 18, 1981), p. 10.

19. William Kincade, "U.S. Targeting Plans," Arms Control Today, Sept. 1980.

20. Keith B. Payne, Nuclear Deterrence in U.S.-Soviet Relations (Boulder, Colorado: Westview Press, 1982), p. 194. He argues that "lacking the capability to defend the American homeland, an American president could not responsibly enter even into counterforce-oriented exchanges with the Soviet Union, both because of the lack of an effective American warfighting capability and because of the likelihood that the U.S. could not survive a resulting escalation process." On the other hand, it is unlikely that an American president would not choose to retaliate in some way after a Soviet countersilo attack that killed 2 to 22 million Americans.

21. Casper Weinberger, Department of Defense Annual Report to the Congress, Fiscal Year 1983, (Wash. DC: GPO, Feb. 8, 1982), pp. 1-18.

22. Keith B. Payne, "Space-based Laser BMD: Strategic Policy and the ABM Treaty", Information Series Number 115, National Institute for Public Policy, July 1982, p. 4. See also, Richard Halloran, "Pentagon Draws Up First Strategy for Fighting a Long Nuclear War," New York Times, Sunday, May 30, 1982.

23. Halloran, Ibid.

24. Ibid.

25. Ibid.

26. Ibid.

27. Jack Kangas, "Strategic Nuclear Policy and Space-based Lasers," unpublished paper, National Institute for Public Policy, April 1982.

28. Most of this information is drawn from Bhupendra Jasani, ed., Outer Space-A New Dimension in the Arms Race published for SIPRI (London: Taylor and Francis Ltd., 1982).

29. Albert Carnesale, "Ballistic Missile Defense: Updating the Debate," U.S. Arms Control Objectives and the Implications for Ballistic Missile Defense, proceeding of a symposium held at the Center for Science and International Affairs, Harvard University, Nov. 1-2, 1979, p. 28.

30. Inis Claude, The Changing United Nations (New York, 1967), p. 9.

6 The Strategic Nuclear Policy of the Reagan Administration: Trends, Problems, and the Potential Relevance of Space-Based Laser Weapons

Colin S. Gray

REAGAN'S INHERITANCE AND THE QUESTION OF STRATEGY

Many observers have remarked that President Reagan's national security bureaucracy did not rush, in 1981 (in time for presentation in Secretary Weinberger's DoD Annual Report for FY 1983), to invent and polish a new-sounding nuclear strategy concept or slogan. To date, the closest the Administration has come to developing general concepts or slogans are the ideas of the "margin of safety" and "prevailing with pride." In common with familiar previous slogans, such as "a defense posture second to none," the "margin of safety" usefully defies precise definition and provides no certain target for criticism. The idea of "prevailing," with or without pride, has proved to be rather more controversial because it has played to the anxieties of those who suspect and allege that the Reagan Administration is planning for the United States to win, or prevail in, a (protracted) nuclear war.

It is an important and considerably under-recognized fact that Secretary Weinberger's DoD has felt no pressing need to redesign fundamental U.S. nuclear strategy. The Reagan Administration, in large part, is conducting the <u>implementation</u> of a strategic vision already substantially delineated by officials in the last years of the Carter Administration. However, although there is a major degree of continuity in strategic policy assumptions and approach that unites Caspar Weinberger's DoD with the DoD of his immediate predecessor, Harold Brown, it would probably be a mistake to attribute that apparent continuity unduly to a meeting of minds.

Mr. Weinberger's DoD has not given the impression of being strong on the consideration of overall strategy questions. This impression and possible fact, reflects the character, inclination and particular talents of top management in DoD, and the official belief that what is needed now are real programs rather than further study and the weaving of intellectualized propositions into yet more strategic doctrine. The handful of senior DoD officials who have credentials as strategic thinkers have tended to work at and promote their more personal defense concerns. This phenomenon has resulted in DoD designing unusually serious policy stories in the realms of civil defense, industrial mobilization, and -- contrary to appearances, perhaps -- strategic arms control.

With respect to military strategy in general, as opposed to nuclear strategy more specifically, Mr. Weinberger's DoD has designed a genuinely global cast to planning for geographical force allocation. Motivated laudably by a determination to be able to regain the initiative in the context of crises authored by others and of anticipated short-falls in U.S. and U.S.-Allied defense capability, the Administration has stressed flexibility in potential force application and the possibility of "horizontal escalation." Analogically, Mr. Weinberger has spoken of considering conventional force employment in terms very similar to those used by John Foster Dulles in 1954 when he said "[t]he basic decision was to depend primarily upon a capacity to retaliate, instantly, by means and at places of our choosing."

A factor that has disinclined the administration from issuing a new strategic-nuclear doctrinal *démarche* has been the felt political need not to fuel further the evident public unease over official nuclear policy thinking. In the first major speech on nuclear strategy to be delivered by a cabinet level official of the Reagan Administration, then Secretary of State Haig (on April 6, 1982) went almost to considerable lengths to point to the continuity in thinking which unites the present Administration to all previous U.S. Administrations since 1945.

In the words of Secretary of State Haig,

> [f]rom the earliest days of the postwar era, America's leaders recognized the only nuclear strategy consistent with our values and our survival.....is the strategy of deterrence. The massive destructive power of these weapons precludes their serving any other purpose.

However, some features of 1970's thinking clearly have been rejected. Secretary Haig offered the opinion that

> [t]he dynamic nature of the Soviet nuclear buildup demonstrates that the Soviet leaders do not believe in the concept of 'sufficiency'. They are not likely to be deterred by a force based upon it.

While endeavoring to deny the value of ideas such as "sufficiency" and "essential equivalence" (a Carter Administration buzz word rationalizing a growing imbalance that was believed to be tolerable, in effect a concept of "tolerable inferiority"), the Reagan Administration appeared to understand that there were more potential political pitfalls than there are potential political benefits in embracing publicly a new, or seemingly new, strategic nuclear conceptual "grand design".

Pulling in the opposite direction, on the other hand, was the felt pressure to provide a coherent policy rationale for the major force modernization and buildup program that had been announced already. The Administration felt vulnerable to the charge that it was simply throwing a great deal of money at a set of ill-defined strategic problems. It is a matter of public record that the administration has not offered a well-crafted detailed explanation of its strategic policy thinking. The President's announcement on October 2, 1981 of his strategic forces' modernization program, and Secretary Weinberger's first Annual Report, for FY 1983, were long on the detail of programs, and no less short on strategic policy rationale. The appearance of a neglect of strategic policy considerations was corrected in 1983 when *The New York Times* and *The Los Angeles Times* leaked what were claimed to be the details of the Defense Guidance document and of the so-called strategic-forces "master plan." Unfortunately, the political benefit that might have accrued as a result of being seen to have strategy effectively was negated by the fact that both of the major news stories in question characterized the Administration's nuclear policy thinking as a radical departure from past endeavors -- towards planning for victory in long, or protracted, nuclear war.

The Reagan election story was to the effect that the United States needed more military power of all kinds, *ergo*, the problem was to rebuild America's military power, rather than to design clever-sounding policy. Secretary of Defense Weinberger came close to saying just this in his Annual Report for 1983.

> [e]ven though it is essential that we reform our policy, one must not regard this reform as a substitute for an increased defense

> effort. The adoption of new ideas and thinking is sometimes presented as an alternative to sustained growth in the defense budget. It is not. Part of the needed reform in strategic thinking is precisely the new realization that we must devote more resources to defense.

Unfortunately for the Administration, the deepening economic recession of 1981-83, the absence of a genuine "mobilization spirit" (as, for example, in 1950-51), the sophistication of some congressional skeptics, and an upswell of public protest activity combined to pressure the government into designing a strategic policy response to criticism. Short of a crisis of a "war is in sight" character, no administration can, in effect, say "let us purchase the weapons now, and calculate their strategic utility in support of national policy later."

Criticism from such Senators as Sam Nunn and Gary Hart that the administration's programs were unduly light on doctrinal or policy rationale, also has been reflected in some problems that surfaced within the official defense bureaucracy. If doctrine and strategy guidance is weak, officials do not know what forces and other capabilities to develop and acquire. Contemporary examples of this difficulty abound. For example, "working level" officials, in and out of uniform, have been told that U.S. strategic forces must have endurance and must lend themselves to flexibility in employment. But, how much endurance is required, for which elements, or fraction of elements, of the strategic force, for how long and for what purposes? Similarly, this administration believes that a prompt hard-target kill strike-back capability is essential for deterrence stability (i.e., the Soviets may strike first, but they <u>lose</u> the counterforce exchange and, given the Soviet payload distribution among fixed and mobile launchers, they lose very badly indeed). But how much hard-target kill capability is needed, and does that capability have to be very promptly available at all times? These are not academic questions. They speak directly to the degree of interest the United States should demonstrate in many individual strategic programs.

In practice, the Reagan Administration has been functioning more rationally than many critics will allow. For example, the annual Defense Guidance document attempts rigorously to map required dimensions for programmed forces and, above all, to point to gaps in programmed forces for missions that cannot easily be eschewed. Similarly, the Administration is seeking to improve on the performance of the Carter Administration, which developed "the policy," witness "the countervailing strategy" and the Rapid Deployment Force, but neglected to plan for the means to implement the policy. The Reagan Administration may (though a measure of skepticism is certainly appropriate) be able to reduce the dimensions of at least two major problems, policy clarity and policy "follow up." A strikingly persistent feature of U.S. strategic policy, so-called, in the 1960s and 1970s, with respect to at least three truly major instances, was that "high level" policy determination, as solemnly reflected in presidential-level documents, was not implemented at all vigorously.

In 1961-62 the U.S. Government decided that it required a far greater degree of flexibility in its strategic nuclear war planning. This policy determination was reflected explicitly in documents and in public speeches by the Secretary of Defense. However, some of the flexibility envisaged by very senior officials in 1961-62 did not actually occur until the mid-1970s. Why? Because, Secretary of Defense McNamara grew cold to the idea. He came to doubt its strategic wisdom and he feared its budgetary implications and he was nervous of public political acceptability, and the Air Force opposed it. However, had there been a proper system for policy implementation, notwithstanding the grounds for official second-thoughts first cited, at least, obstruction and disinterest (not to mention the reassignment of key individuals) could not have played their parts to abort what was, and remains, a sound idea.

On January 17, 1976, President Nixon approved NSDM-242. This authoritative document (which was to be superceded by PD-59 of July 25, 1980) called for a very considerable expansion of the number of targeting option packages, emphasized escalation control, a secure strategic reserve force, and the need to thwart the ability of the Soviet economy to recover from nuclear war. NSDM-242 was implemented only with partial success. Key individuals were removed from office or were reassigned. Some of the strategic missions were impossibly difficult to plan for. For example, how could the JSTPS target "the Soviet recovery economy," and what <u>was</u> "the recovery economy"? The controlled use of strategic nuclear forces envisaged almost as the centerpiece of NSDM-242, was intolerably stressful of available and programmed U.S. C^3I assets. There can be little political bargaining value to a "secure strategic reserve force," if that force is beyond communication reach, cannot be commanded, and has no intelligence gathering and analyzing capability available to inform the targeting.

PD-59 of 1980 had its origins back in 1977 in the Presidential policy document PD-18. PD-18 required studies of targeting policy; of the proper character, functions, and necessary supporting C^3I for the secure strategic reserve force; and of U.S. counterforce needs. The targeting review, colloquially known as the "Sloss Report" (after Leon Sloss, who headed the targeting review effort under Secretary of Defense Harold Brown), substantially was completed by the end of 1978. Some of its recommendations received President Carter's blessing on July 25, 1980 in PD-59.

The principal and most obvious features of PD-59 were the requirements that U.S. strategy address itself explicitly to defeat Soviet strategy, hence the public emphasis that came to be placed upon denying the U.S.S.R. any plausible prospect of achieving victory; a shift of emphasis toward targeting Soviet military forces and political-military leadership cadres and facilities more vigorously than had been the case before; a renewed and augmented emphasis upon developing plans and capabilities for the waging of protracted war; a partial redefinition of the character and support requirements for a secure strategic reserve force; substantial abandonment of one of the major innovations introduced via NSDM-242, the requirement to hold at risk the Soviet ability to effect prompt economic recovery from a nuclear war, a concept replaced by the requirement to threaten Soviet logistics and war-supporting industry; and, finally, PD-59 looked to the flexible, interchangeable (where technically feasible) and complementary employment of theater and intercontinental nuclear strike forces.

It is possibly unfair to comment at all upon the policy performance of the Carter Adminstration with respect to PD-59, given that PD-59 was signed only six months prior to Mr. Carter's departure from office. Nonetheless, it is fair to note that the Administration was sufficiently ambivalent over the content of the "Sloss Report" that PD-59 post-dated its base document by all of eighteen months. Furthermore, even had key Department of Defense officials not strayed from the plain intent of the authors of PD-59, where were the effecting defense programs? Where were the C^3I assets that could survive, be reconstituted, and endure for the conduct of a protracted war? Where was the civil defense program without which the United States as a society would not survive a protracted war, even if its forces could?

Notwithstanding the criticisms advanced above, a coherent policy, like Rome, cannot be built in a day. Unfortunately, a president, or an NSC staffer, can "make policy" in a declaratory sense in an afternoon; it may take a decade or longer for the "policy" to be reflected at all adequately in physical capabilities; by which time, two or more administrations later, declaratory policy and policy guidance doubtless will have altered considerably. It has long been observed that "policy" has three components: declaration, capabilities, and action. When talking of trends in policy, as here, people should never forget that words, in and of themselves, spoken or written, do not suffice to a policy make.

SPACE WEAPONRY AND THE POLICY DEBATE

The point needs making at this juncture that at the present time the United States Government neither provides policy guidance with explicit reference to space-based laser weapons, nor has it designed any important aspects of its "defense guidance" with the availability of space-based laser weapons in mind. These conditions are likely to change in the near future. However, this study explicitly has a current focus. The purpose here is to explain the trends in nuclear policy thinking: to establish the policy context into which consideration of laser weapons for a variety of strategic tasks may fit. The military space technology community today must cope with the policy world as described and analyzed here. The potential relevance of space-based lasers to effect or support the clearly identifiable trends in strategic policy is developed later in this chapter. However, the truly intriguing question, is whether the technological promise of space-based lasers is sufficiently attractive that the United States Government should alter its strategic policy to accommodate the new techology.

The germ of the case for deep and broad-fronted exploitation of space-based laser technology already is present in the official deterrence policy story. As the U.S. Government comes to understand that collateral damage is the mortal foe of escalation control, and as it embraces even more wholeheartedly than is the case today a doctrine of continuing deterrence for prolonged conflict, the ability physically to limit (not preclude, necessarily) damage to the American homeland, must increase and to have alternatives to inflicting damage on Soviet territory, the attractiveness in principle of multi-layered active and passive defense must rise.

Whether the U.S. policy debate over the merits of space-based laser deployment comes to be joined as a consequence of some mix of maturing technical possibilities, some Soviet-authored triggering event or events (a Soviet "demonstration" *à la Sputnik*, for example), or political pressures from strong believers, the familiar broad liberal versus conservative debate is unlikely to reappear in anything like its pure form. Many conservative, defense-minded officals and commentators, who have no generic doctrinal objection to laser weaponry, will worry over, and possibly argue about the prospect of a contest in space that may not be resolved to the advantage of the U.S. Some very senior Reagan Administration officials already have begun to wonder whether there might not be some disturbing functionally analogous aspects linking conflict in space to conflict at sea. War at sea, theoretically, could be waged in preference to war on land, because of the discreteness of the target sets, the near, though not quite total, absence even of the possibility of collateral damage, and in general, because of the relative ease of crisis management and escalation control. Unfortunately for the United States, there is a structural, geopolitically based asymmetry to the Western disadvantage in war at sea. Specifically, the West has a far greater strategic need to command and use the sea than does the U.S.S.R. Indeed, the Soviet Navy needs not so much to enjoy freedom of action at sea, but rather to deny the free use of the sea to the maritime alliance of the West.

Imperfect, and possibly even inappropriate, though the analogy may be, this analogy is in people's minds. If, on balance, the United States is likely to remain more dependent upon assets of all kinds in space than is the Soviet Union, then how can space as a new fourth dimension of potential warfare possibly work to the West's advantage? If this line of thought is correct, it will follow that achievement even of a stalemate or stand-off in space may function to the Soviet advantage. For reasons that do not belong here, this author believes that the line of argument developed immediately above, although not totally devoid of merit, is both, on balance, in error and misleading, and above all is irrelevant. In the same way that armed conflict at sea cannot be precluded by conscious international legal design, neither can conflict in space. The very fact that the United States has a large and very rapidly growing C^3I investment in space serves to insure accelerated Soviet interest in contesting freedom of activity in the fourth dimension. Given the substantially irreversible scale of the U.S. commitment to develop and deploy military space-based C^3I assets, the clear trend toward potential conflict in space similarly is irreversible.*

These introductory comments on space-based weapon issues are deployed as scene-setters for the discussion below of current trends in strategic policy. Scene-setting in this terse manner is appropriate because the government officials who are the authors and implementors of contemporary strategic policy have yet to confront the truly difficult and challenging problems pertinent to the design funding and implementation of a military space policy. It is scarcely an exaggeration to observe that space weaponry, for ASAT, BMD, or air-defense missions, for the pertinent examples, currently is a slightly off-stage set of "wild cards" that all strategic policy debaters know may or may not be dramatically supportive or destructive of the premises that underpin policy design and programs today. Moreover, everybody knows, but need not yet face the prospect, that space weaponry may come to effect a genuine revolution in warfare, in particular with reference to the relationship between the offense and the defense.

Visionaries, romantics, and technical enthusiasts, have a wide range of more or less well-considered "space war" stories to tell, but sober, pragmatic and professionally rather unimaginative defense analysts, who tend to discount strategic "vision" assuredly will not confront the strategic revolutionary possibilities inherent in military space technology until they have to. History tells us that very few new weapon technologies have a "war-winning" quality about them during their technical infancy and adolescence, and even thereafter wars tend to be won by the efforts of many arms employed properly in combination. It is far more likely than not that space-based laser weapons, out to the end of this century, will function much as tanks did in the last two years of World War I. The tanks of 1917-18 were not, and could not be, war-winners, but they were extremely useful when sensibly employed by forces who know their strengths and limitations. The cause of the tank was not much helped by "visionaries" who, virtually on the backs of envelopes, created self-sufficient tank armies which could, so it was argued, supercede other kinds of forces. A similar phenomenon already has appeared with reference to space-based laser weapons. A possibly very useful weapon is being touted by some as a likely war-winner on a predicted time scale that is likely to discredit the entire enterprise.

---

* See Colin S. Gray, U.S. Military Space Policy to the Year 2000 (Cambridge, Mass.: Abt Books, 1983).

The Reagan Administration is being lobbied by groups of "space war" visionaries, and the net effect of this pressure is to retard the balanced policy consideration of near and medium-term achieveable space weapon capabilities. The Reagan officials responsible for the policy trends detailed below are, to some important degree, are not assisted in their duty to make timely decisions on military space policy by the extreme character of the technical and policy claims advanced by rival groups of space weapon proponents and detractors. The policy and technical debate over space weaponry of all kinds, not only lasers, is directly relevant to the evolution of strategic policy. But the average of high-level policy- makers tend not to be well-equipped technically to pass judgment between rival claimed certainties. As with all defense policy issues, space weaponry will be developed and deployed as a consequence of the playing out of the policy process and will be impeded and advanced by bureaucratic vested interests.

NUCLEAR POLICY AND PROGRAM TRENDS UNDER PRESIDENT REAGAN

As noted earlier, the Reagan Administration inherited a strategic policy, in a conceptual sense, with which it found little fault. The perceived challenge essentially was to translate the "countervailing strategy" into appropriate detailed policy guidance, and to identify the necessary hardware programs and secure their proper funding. Also, some refinement, clarification, and alteration of program priorities was deemed necessary. What follows in this section are summaries of the current trends in official thinking on strategy and on the programs needed to implement the strategy. Some, if not most, of this discussion will be familiar from the earlier brief presentation of the evolution of strategic thinking in the 1970s and does, of course, reflect the truth that at the broad conceptual level of approaching issues of deterrence stability, Harold Brown and Casper Weinberger inhabit the same universe.

1. The window of vulnerability and quick fixes. Quite aside from issues of strategy, a preeminent concern of the administration is to do what can be done rapidly. Defense policy is the art of the possible. Some of the more desirable strategic weapon programs, which do speak to the most serious American deficiencies, simply cannot be accelerated so as to provide near-term desirable strategic capability. For example, a sur-

vivably-based MX missile force is needed almost desperately for leverage in START, and to strengthen deterrence stability. However, for reasons in part of its own making, the Reagan Administration has no ICBM (let alone MX-ICBM) quick-fix options politically available. The quick-fix for the augmentation of non-survivable firepower offered by the President, to place perhaps the first three years of MX missile production in Minuteman III silos was rejected categorically by the Senate Armed Services Committee. The other reliable, if possibly delayed-in-execution, hard-target threatening program favored by the administration, the Trident D-5 SLBM, has an IOC planned at present for 1988/89.

To note that the Administration has placed highest priority upon the modernization of the manned bomber force, upon producing and deploying the whole family of cruise missiles (ALCM, SLCM, GLCM), and upon upgrading C^3I to that prompt retaliation will always be assured. The C^3I modernization program is not focussed upon investment in systems that could endure for several weeks or months. However, these top priorities chose themselves for reason of their relatively near-term availability.

The plan to acquire 100 B1-Bs, to re-engine the KC-135s, to acquire more KC-10s, and to provide all the B-52 G's and H's with a new Offensive Avionics Systems (OAS) as well as to give some B-52 H's EMP protection, does not signal a shift in strategic thinking away from ICBMs. In effect, the administration asked itself what it could do in a hurry that would not be retarded by bitter political battles and would not entail taking major technical risks -- the answer was to modernize the bomber force. Similar reasoning applies to many C^3I programs and certainly to the acquisition of strategic cruise missiles.

The administration realized that the United States faced a severe strategic-force survivability problem through the 1980s; two-and- a-half legs of the triad would be vulnerable to prompt destruction in a Soviet surprise attack. Minuteman-Titan, to all intents and purposes, is totally vulnerable today; a bare handful of bombers could escape with high assurance (in fact the number of alert B-52s is lower today than it has been for many years); and one-third to a half of the SSBNs could be destroyed in port. Traditional and very familiar arguments concerning the "synergism for survival" that the ICBMs and the bombers provide for each other is now subject to considerable official question because of the Soviet "pin-down" threat. In theory, the Soviet Union cannot attack the ICBMs *and* the bomber bases without providing the warning time for one or both (if the first echelon of the attack is a salvo firing of ICBMs in response to which the United States launches some of its ICBMs on warning) to flush and escape. Technical and tactical competence on the Soviet part may already have placed this long-standing survival theory in jeopardy. Soviet SLBMs could assault the bomber bases and pin-down some of the *Minuteman* force.

The U.S. Government argues, in response, that as the number of Soviet SSBNs patrolling from viable firing positions alters, then the alert status of U.S. strategic forces is changed. Nonetheless, there are gounds for legitimate concern in this area. Should the United States decide to live indefinitely with all or a noteworthy number of its ICBMs dependent upon LUA for survival, the ability to thwart an SLBM pin-down attack will be of growing importance. This is the kind of strategic mission for which BMD of several kinds are almost ideally suited. The attack would be relatively small and would not be of the large salvo kind.

2. *A "warfighting" doctrine of continuing deterrence.* The Reagan Administration believes that although a central nuclear war cannot be won in any manner compatible with American cultural values, a nuclear war can be waged in such a manner that the enemy's war plan is thwarted. Its strategy would be defeated. Victory denial capability, which is not to be confused with victory, provides robust support for deterrence stability because it is calibrated, as best as can be achieved, by constant reference to adversary thinking, plans, objectives, and capabilities.

   The administration employs the concept of deterrence as it should be employed, with reference to an end result in the minds of potential enemies, and not with reference to a particular strategy or to particular kinds of forces. Official Washington, today, does not acknowledge a distinction between warfighting and deterrence. The Carter Administration, with the "countervailing strategy" and PD-59 agreed, in principle, but in reality there was a great deal of reluctance in high places to embrace the defense program implications of the shift in emphasis in authoritative official strategic doctrine.

Mr. Weinberger's Defense Department is not at all ashamed to admit that the MX and eventually the _Trident_ D-5 have value very substantially because they have a major impressive capability to threaten and destroy Soviet hard-targets. 1981-83 has seen a significant shift in the extant official trend towards acknowledging the necessity for acquiring strategic forces which can combat Soviet forces directly.

3. _A revival of interest in strategic defense._ The administration, hardly surprisingly, does not speak with one voice on the subject of the desirable balance of effort to be invested in the strategic defense as opposed to the strategic offense. At issue here is both a matter of strategic-intellectual, or doctrinal, education, and a major question of technical and political-social feasibility. Many senior Reagan officials have no doctrinal quarrel with the proposition that it is more important to defend Americans than it is to threaten or kill Russians. Their skepticism appears to lie almost wholly in the region of technical plausibility. Can Americans be protected? If so, how many? Can the Soviet Union move to offset U.S. active and passive defense measures?

Plainly, there is a new trend toward greater provision for active and passive defense, but that trend today is in its infancy. Indeed, with respect to the Reagan plan to invest in civil defense over the next five years, the trend has been substantially aborted by the incredulity of the Congress. There has been public talk by senior defense officials of a "revolutionary" shift in U.S. policy direction towards the defense, but that talk is not matched by currently planned programs. Serious though the proposed civil defense programs appeared to be at $4 billion, the Administration was not sufficiently committed to it even to attempt to offer very vigorous policy justification.

The logic of the authoritative official policy story on continuing deterrence should lead to strong advocacy for the provision of a multi-layering of strategic capabilities, offensive and defensive (active and passive) in character, for the goal of damage-limitation. The administration, to date, has not engaged in such advocacy. Possible reasons for this neglect include a failure to comprehend the true dimension of the strategic program that their doctrine of deterrence really requires, or an unwillingness to confront nuclear policy issues, head on, in the current national and international climate of opinion. The defense story has, in practice, been told piecemeal, with the predictable result that none of the pieces, considered in isolation, looks very credible. For example, the revitalization of North America's air defense is difficult to justify in the absence of a persuasive rationale concerning provisions to protect against missile attack. Similarly, most strong claims for the protection that might be secured through civil defense virtually invite skepticism, if not outright derision, if the damage-limitation argument is confined to civil defense.

To date, the Administration is interested in BMD, but is vastly unresolved in its official mind as to the net value to the United States of abandoning what many policy-makers see, in effect, as the security blanket of the ABM Treaty. Civil defense is proposed at a serious funding level, but the administration has given evidence of being either unwilling or unable to attempt to make a vigorous policy case in its defense in the face of a great deal of political hostility and, to be fair, genuine and technical skepticism. Mr. Reagan's proposed civil defense program obviously was going to be controversial.

Because some key senior members of Mr. Reagan's defense bureaucracy clearly do appreciate the synergistic value of a balanced offense-defense national strategic posture, it need not follow that the administration either at the "working level" or at the cabinet and immediate sub-cabinet level shares that understanding. The most that can be said concerning official commitment to the achievement of a more genuine balance between the offense and the defense is that the trend is towards a better balance, but the trend is at such an early stage of manifestation that confident prediction of how far it will proceed must be strictly speculative. The strategic style of the current national security policy leadership group has to discipline any tendency towards interpreting the evidence of individual defense programs as signs of a grand design for a coherent strategic package intended to provide a non-marginal capability to limit damage. To date, the full logic of the official deterrence theory has yet to be articulated at the most senior level of government. Former Secretary Haig has spoken of the strategic offensive requirements of his theory of deterrence, but he has not attempted, in more than a cursory way, to relate the credibility of threatened offensive action to a perceived ability to defend the United States to an important degree. While there is evidence of new-found interest in defense, the verdict has to be "not proven" as yet with regard to the measure of official commitment to homeland protection.

4. <u>A major commitment to the survival and endurance of strategic forces, to the continuity of government, and to the survivability of C^3I.</u> Notwithstanding the presidential-level policy statement of NSDM-242, and the subsequent policy guidance provided in the NUWEP document, very high-level official American commitment to the idea of the controlled sequencing of nuclear strikes has not, to date, been capable of operational implementation within the realm of physical practicality. U.S. strategic nuclear forces and their supporting C^3I have been postured for spasm war.

NSDM-242 called for the survivability and endurability of strategic forces, C^3I, and a functioning National Command Authority, but it could not create them with a stroke of the pen. In the early 1980s the United States still lacked robustly survivable C^3I. The Reagan Administration inherited a defense community-wide understanding of the urgent need to invest in C^3I architecture that could survive an attack, preferably that could survive several attacks, and that could gather, transmit, and analyze sufficient information for the surviving strategic forces to be able to continue their deterrent function. A vital part of that architecture, of course, had to be a survivable, functioning NCA that had enduring access both to processed information and to the forces to be commanded.

The Administration believes that the ability of the United States to conduct a protracted (nuclear) war is important for deterrence, and is simply prudent given the very uncertain character and duration of a general East-West military struggle. U.S. protracted war capability should strengthen Soviet perception of the U.S. as a very strong opponent, which must be a plus for deterrence, in that it has to discipline Soviet predictions as to the prospects for achieving swift military victory. For several years now, certainly well prior to the signing of PD-59 in July 1980, some American defense analysts have believed that the Soviet leadership is particularly fearful of the political and social risks that they would run in an armed conflict that could not be concluded swiftly and victoriously. The sacred texts of Marxism-Leninism, and an overwhelming weight of actual historical evidence, affirm that "war is the midwife of revolution," or at the least, of political change. Although it is distinctly possible that the geopolitically far-flung maritime alliance of the West would suffer from defection and lack of cohesion in the event of protracted war, the Soviets cannot be at all confident that their Empire, in its core area as well as in its outer marches, would not come unravelled under the pressure created by the multi-faceted damage of a prolonged campaign. It is not at all obvious that a protracted (nuclear) war would be a circumstance wherein the Soviet leadership could enact a political replay of Stalin's assumption of the mantle of the defenders of Mother Russia.

Soviet ideology mandates affirmation of the belief that the "stability of the home front" is of vital importance in armed struggle. With American preparedness to wage protracted war comes the promise of placing Soviet imperial control under a unique strain. This is the kind of danger which the Soviets, with their near-paranoid concern for total political control, are most unlikely to dismiss lightly or to discount.

To date, although the Reagan Administration has embraced the general idea of protracted war, the policy story remains distinctly murky. To a degree this is understandable. It is very difficult to specify the military and other related requirements of a protracted war strategy. Also, it is difficult to persuade officials to focus upon protracted war problems when they are very anxious about U.S. ability to hold its own on the first day of a conflict. But, inherent difficulties admitted, decisions need to be made and detailed guidance issued. The Administration has yet to decide whether it means by protracted nuclear war a war which proceeds for weeks and perhaps months and does not see a SIOP-RISOP level exchange of any kind, at least of any major kind; whether it means a war that proceeds after a SIOP-RISOP exchange sequence which is very carefully designed so as to minimize collateral civilian damage; or whether it means a war which proceeds after each side has struck the cities of the other. Different agencies of government have developed different preferences among the alternative assumption-sets specified above. Some elements in the the Air Force, and more generally, the strategic targeting community, for example, are very reluctant to encourage the idea that a nuclear war could proceed for days and weeks, let alone months, without execution of SIOP level attacks. The guardians of the SIOP are understandably nervous of the attrition which U.S. strategic forces, even on-alert forces, would suffer were they required to stand by ready to go for weeks on end in a wartime environment. Political and strategic sense may argue for the withholding of nuclear strike forces, but military efficiency, narrowly defined, would argue for relatively prompt execution of SIOP- level attacks.

The U.S. Government, today, is aware of the problems of the predictable protracted war tension between politico-strategic incentives to go "light and late," and strictly military incentives to go "large and early." This is a comparatively recent problem and it flows from the facts that the U.S. strategic nuclear force posture and its supporting C^3I, as noted above, has been designed for spasm employment, and that posture and doctrine are somewhat out of phase. Ironically, perhaps, U.S. strategic forces and C^3I have grown markedly physically less survivable in the very period when their survivability and operational endurance have assumed real policy importance for the first time.

The Reagan Administration is very serious indeed about the importance of protracted war capability for pre-war deterrence and for the continuation of a deterrence policy in war itself. Official plans for survivable strategic C^3I, and for the continuity of government, genuinely are impressive. The fragility of U.S. C^3I and NCA survivability has, until now, reduced doctrinal debates over the merits and demerits of strategic flexibilty largely to academic insignificance. The Reagan program to modernize and add major redundancies to C^3 architecture necessarily begs many questions of technical detail. For example, U.S. satellites in geostationary orbit are not detectable by Soviet radar today, but will that still be the case in 1995 or 2000? Even if many satellites will remain undetected, can the same be anticipated for their ground, air and sea-based receiver nodes?

The enduring survivability story for the manned bomber/cruise missile carrying-force is reasonable, always provided one disallows a range of surprise attack considerations. The SSBN story for protracted war is a strong one, provided one assumes an initial generated alert condition and provided one chooses not to worry unduly about the sharp decline in the number of hulls in the 1990's. Without forgetting the C^3 problems of submarines, the MIRV "footprint" issue that impacts on the extent of the sea-room that is compatible with the high state of readiness on station, the _Trident_ D-5 fits well into the protracted war concept. The proliferation of TLAM/Ns on large numbers of surface and subsurface platforms will compensate, to a limited though important degree, for the concentration of nuclear firepower implicit in the size and small number of _Ohio_ class SSBNs.

The major missing element in the policy story of protracted war pertains to the enduring survivability of _people_. It is probably unreasonable to seek to insist that civilians be rendered as "hard" as the weapons that defend them, but it is time the government noticed the fact that although a strategically useful fraction of the U.S. strategic nuclear forces should endure, as should a functioning NCA, given current programs, the country, in the essential sense of the civilian population, would not endure. Early in 1982 the Administration was challenged to defend its civil defense program, and to explain the optimistic-sounding claims issued by T.K. Jones to the effect both that the digging of "hasty shelters" could, and almost certainly would, save millions of lives, and that the United States could recover from a large-scale nuclear war in 2-4 years. The challenge was not accepted. This episode was doubly unfortunate in that it may have "poisoned the well" for civil defense program justification for a long time to come, and it left a gaping and truly fatal hole in the entire protracted war thesis.

Some key individuals in the Department of Defense certainly understand that the "continuing deterrence" doctrine really is nonsense in the absence of a major civil defense program. Such absence, asymmetrically between the superpowers to the American disadvantage, will grant the Soviet Union a very important instrument for crisis-time coercion (urban evacuation), and will leave American aspirations for control and success in (and hence deterrence of) protracted nuclear war unbearably dependent upon the exercise of extreme targeting restraint on the Soviet part.

5. Flexibility in targeting options. The virtues of flexibility in strategic nuclear war planning have been well appreciated in principle by senior civilian policy-makers since February 1961, at least. At that time Secretary of Defense McNamara paid a memorable visit to the Joint Strategic Target Planning Center at Offutt AFB and apparently was shaken and appalled by the inflexibility of the extant SIOP, which had as its *leitmotiv* the concept of the "optimum mix." NSDM-242 of 1974 established at a high level the political requirement for flexibility for escalation control, and the JSTPS responded as best it was able over the next several years. Strategic flexibility has been endorsed because a degree of choice in stratgic employment long has been judged to be inherently desirable for credibility in deterrent threats and as a means for escalation control; because new weapon and weapon-related technologies permitted greater missile accuracy, rapid missile retargeting (with the Command Data Buffer System), and higher yield to weight ratios in warheads; and because it has been hoped that refinements in employment policy in declaration and in actual plans, could function as a "strategy offset" for the decline in relative strategic muscle.

Unfortunately, as noted above, strategic flexibility in controlled employment was always more aspiration than serious promise through the 1970s, for reason of the vulnerability of U.S. C^3I. A series of studies in the middle and late 1970s on strategic command and control demonstrated beyond reasonable doubt that U.S. strategic forces were not postured to endure, even if they could survive a first-strike; strategic C^3I would be hard pressed to survive attack even to the level simply of providing for operational connectivities in a very austere mode; and that enduring NCA connectivity with forces and such C^3I assets as survived would be problematic at best. In short, the United States had constructed a strategic nuclear war machine that was well designed to fulfill its long-standing mission(s), of going first in a very large way, and was probably adequate for retaliating promptly and massively.

Somewhat more specifically, U.S. strategic planners in the middle and late 1970s established a secure strategic reserve force which was not, by definition, committed to SIOP missions, and which was not even pre-targeted. It was the very expression of flexibility for political insurance in the end game or aftermath of a major war. However, the weaknesses in U.S. strategic C^3I, including NCA survivability and connectivity, virtually made nonsense of the idea of a secure strategic reserve force. Notwithstanding the authority of NSDM-242 and its implementing NUWEP guidance papers, some (and perhaps even very many) senior civilian officials in the 1970's harbored deep strategic misgivings, in principle, about the strategic virtues of great flexibility in employment planning and use capabilities. Some officials worried that strategic flexibility would function as a justification for service "requirements" for virtually open ended procurement of warfighting hardware; would stimulate the arms race needlessly; and might mislead a president into believing that the controlled nuclear wars of the strategic planners was some approximation to likely reality.

In good measure it is accurate to claim that Nixon, Ford and even Carter defense policy-makers sought to separate changes in strategy, as reflected, more or less closely, in war planning, from procurement issues. The result, inevitably, was an evolving force posture that was not well suited to perform its changing mix of strategic mission options. The Reagan Administration is endeavoring to correct this now-traditional mismatch between strategy plans and capability. Similarly, this Administration, following Harold Brown's example, also is seeking to achieve a reasonably close match between declaratory policy and operational strategy.

Protracted war doctrine, as well as common sense strategic logic (of a kind familiar for more than twenty years), mandates major investment in the hardware and software that provides targeting flexibility over time. ICBMs and bombers that cannot survive a surprise attack cannot be employed flexibly. Either they are not available to be employed, or they have to be rushed into action in a "use it or lose it" mode. The ideal that the authors of PD-59 harbored as a distant aspiration was a strategic force posture so stable in its survivability, and in the survivability of its C^3I, that it could be wielded in major or minor key to meet the needs of the real-time political and strategic situation that would evolve over the course of a campaign that might occupy weeks or even months. It is unlikely that strategic forces ever can approach this ideal condition, if only for reason of the technical synergism for reliable mission performance which mandates that particular attacks must be of a certain size and character. Nonetheless, the trend is important, if only as an aspiration. That trend is way from "SIOP thinking" which, _in extremis_, entails the massive, well-coordinated, spasm or "spike" lay down.

The Reagan Administration is well aware of the fact that its protracted war thesis (for deterrence) requires strategic flexibility. Strategic forces to be employed three weeks or three months after the onset of hostilities must survive physically; must endure in an operationally ready status; must be connected to political authority; and must be instructed what, where, and when to strike. This vision of strategic requirements is very stressful of C^3I. With respect to a protracted conflict, the government must assume that its C^3I support structure for the strategic forces will be subject to discrete, though very determined, attack.

It is probably worth mentioning the point that some defense officials today wonder whether strategic flexibility is not one of those good ideas which, when taken too far, becomes a bad idea. A president may be given too much strategic operational choice. He may not understand the full character and implications of the range of options available to him, he may suffer from _un embarras de richesse_. In addition, a great measure of flexibility in planning may well lead to, or at least encourage, the withholding, deployment, employment and even the acquisition of forces, in ways and of kinds which affront military logic.

The limited (and regional) nuclear options of a sub-SIOP character that Secretary of Defense James Schlesinger chose to emphasize in 1974 in his public explanation of NSDM-242, are very considerably less popular today than they were then. The Reagan Administration appears to understand that although sub-SIOP level nuclear options (SNOs) fit very conveniently into a policy story of protracted nuclear war, far more conveniently than does a major "spike" counterforce exchange, they have no known analogue in Soviet military thought or operational planning and they are the tactic of the strong, not of the relatively weak. Preeminent among the strategic nuclear targeting concerns of the Reagan defense community is the determination to defeat a Soviet counterforce attack; pending the availability of the MX ICBM and the *Trident* D-5 SLBM, this mission can be attempted, at least in a relatively slow-motion manner, by penetrating air breathing forces.

6. *A more convincing launch-under-attack (LUA) capability.* The Reagan Administration has not chosen to favor LUA as *the* survival (for prompt strike-back) theory for ICBMs, but many officials fear lest LUA may be the only available theory for ICBM survivability. In addition, the growing threat to manned bombers on runways and during their fly out is prompting serious reconsideration of ways in which the threat to SAC air bases can be diminished, delayed, or evaded.

It is exceedingly unlikely that Congress will permit MX basing in *Minuteman* silos, though the slim possibility cannot be discounted totally. Suffice it to say that there is, at present, an unusual degree of official interest in the enhancement of the technical and political capability for LUA. Some critical elements of the C^3I modernization program could be invaluable for the credibility of LUA options, as well as being essential for the support of surviving and operationally enduring strategic forces in a protracted conflict. For most of the 1980s the United States will have a second strike ICBM force *only* if LUA is the firing tactic. LUA is a dangerous operational policy, but it may be the only policy that fits the near-term strategic context. LUA is vulnerable to the severe degradation of warning and attack assessment sensors, even to brief political irresolution by the NCA, and to a series of careful pin-down strikes.

7. <u>A better integration of defense and arms control policy as a component of overall national security policy.</u> It may be misleading to cite as a trend the growing recognition of the actual and possible relations among defense, arms control, international economic, and more general foreign policy. In a well ordered and sophisticated policy-making universe, such integration would be second nature to all the bureaucratic players. In part for reason of the structure of the U.S. Government bureaucracy, which comprises semi-sovereign entities defending their own parochial interests, the connections between different streams of external policy all too rarely are assured for consideration of their mutual support and compensation.

The grand architecture of U.S. foreign policy can really only be designed and constructed, and then orchestrated, from the White House. Until quite recently, the cast of principal characters in the Reagan Administration's foreign policy "team," and the controls that divided them, virtually precluded any activity worthy of being described as policy planning. A new trend has set in with appointment of "team players" to lead the NSC staff and at the State Department, but the Administration remains noticeably deficient in the realm of "grand strategy" designers. As stated above, the Administration remains very sensitive to the charge that it lacks a coherent policy framework for the guidance of different streams of activity that need close coordination.

Symbolizing and providing some reality to the trend towards greater policy coherence have been the charges at the NSC in the White House. Judge William Clark has more authority than did Richard Allen, and hence has a greater opportunity to orchestrate national security policy from a central standpoint. Important though structure and terms of reference can be, what really matters is how the structure and terms of reference are employed. The new leadership group at the NSC has sought to design a new-looking "grand strategy," but the results to date, leave much to be desired. President Reagan's NSC staff, though of increased importance, continues to be short of numbers of troops and of very senior professionals.

Mr. Reed's and indeed Judge Clark's endeavors as strategic thinkers are unlikley to bear fruit both because there is relatively little manpower supporting the activity of strategic thought and because, at the level of inspiration as opposed to detailed supporting staff work, people with plausible credentials as designers of a grand strategy are not associated with the enterprise of strategy production. Nonetheless, the Administration has recognized the need at least to appear to have an organizing framework for planning diverse threads of foreign policy action, and is very much in the market for strategy, grand or otherwise. It is a general truth that sweeping statements of foreign policy design tend to be written to offer retrospective justification for past decisions, to offer as few hostages to fortune as possible and, overall, to be substantially worthless for policy guidance. However, such documents can provide valuable clues to official thinking, more often than not of a kind unintended by the official authors. Statecraft is an art which can neither be taught, notwithstanding the existence of university-housed schools of diplomacy and so forth, nor planned in any detail. It is a "seat of the striped pants" pragmatic activity that should be guided by some enduring principles specific to the geopolitical interests of the country in question.

Search for a grand strategy is praiseworthy, in that it is healthy for senior policy-makers to contemplate how activity "A" may impinge on activities "B" and "C", while it is virtually mandated by the appreciation of near-term relative U.S. military weakness. The Reagan Administration is aware, in principle, of the fact that it can play political and economic pressure and cooperation cards to attempt to compensate for defense deficiencies and to facilitiate the arms control process. Above all else, perhaps, and this point is vitally important for all major weapon projects, current and near-term future, the Administration has come to appreciate that it has neglected to encourage and sustain the necessary domestic and allied political base for the defense modernization program. Mr. Reagan and his principal advisors realize now that while the American people are generally very suspicious of the Soviet Union, and certainly are willing to have 7 percent of the GNP invested in defense functions, they are not willing to acquiesce in what appears to be an official "blank check" attitude towards the Department of Defense. For many months, the new administration appeared to forget the ancient truth that democracies in peacetime require fairly specific policy justifications for large-scale defense expenditure. It is not sufficient merely to cite "Soviet military power," even in detail; in addition it is necessary to explain how the dollars requested relate to offsetting that power.

The current "freeze" movements in the United States threaten damage to the development of sensible and orderly arms control policy and defense programs. That movement could have been deprived of a great deal of ammunition had the Administration attended properly to developing and explaining the master foreign policy story, and had the U.S. and NATO-allied public been presented with a grand strategic design that placed the limited rearmament program, persuasively, in a persuasive political context.

UNFINISHED BUSINESS

Political and strategic processes are inherently dynamic. It follows that policy in all its aspects similarly must be adjusted in detail, though preferably not in principle, as the problems that it addresses change their form. To claim that there are very important policy issues currently unresolved is not, in and of itself, much of a criticism. There will always be unresolved policy issues. Nonetheless, the issues unresolved in the American case have a tendency to be more serious in kind than are the issues that escape definitive resolution in Moscow. The reason for this is that there is a settled character to Soviet statecraft and strategic thinking and planning that is absent in the United States (or, indeed, in most democracies). The steadiness of the Soviet policy course is not threatened periodically by utopian impulses, whether they stem from an individual leader with a strong "vision", or from an unduly innocent populace, nor does Soviet defense policy have to contend with an important domestic constituency that remains unconvinced that armed forces, even when nuclear-weapon equipped, are necessary.

This chapter has identified and discussed trends. Virtually by definition, trends are incomplete. What follows is a summary of the major items of unfinished business that currently are in strategic policy contention.

1. The overall "integrity" of policy. It is not at all self-evident that President Reagan's programmed defense buildup will constitute a coherent

and appropriate match with stated defense policy guidance (in support of very expansive foreign policy goals). It is easy to criticize, but there are no magic benchmarks that reliably can indicate "how much is enough." Nonetheless, the Administration, to date, has given the impression that it has decided on what to build militarily first, and only subsequently has it begun to devote any careful thought to the strategic missions or foreign policy relevance of the already programmed forces. This criticism is, in reality, more than a little unfair. But an administration which neglects to explain the foreign policy and strategy rationales for its defense program really has only itself to blame when it comes under fire of this kind.

2. The clarification and intepretation of "high policy". As discussed above, many of the strategic ideas that currently enjoy fashionable status among senior policy-makers, are unambiguous enough. But, the Administration has a major task to perform in the vital region of detailed policy guidance. For example, officials endorse the necessity for the United States to invest in a strategic equivalency (or better) in counterforce effect. Does this require promise of counterforce-targetable American weapons capable of denying the Soviet Union victory; of greatly complicating Soviet attack execution; or of defeating the military assets of the Soviet state and, thereby, "winning the war" on the offensive side? Some, though by no means all, senior officials have advertised an historic shift towards the defense in the Reagan strategic program. How radical should this shift be? What are its objectives? Is the principal objective to make the United States appear to be a more robust, a tougher adversary? Or are we witnessing the beginning of a politically purposive trend towards a defense-dominant strategic posture?

The Reagan Administration clearly would like to defend Americans physically. Witness the new, if substantially aborted, civil defense programs, and the modest though significant modernization program for North American air defense. Is the objective to make a strategic policy gesture, putatively of some uncertain deterrent benefit? Or, is the Administration signaling a small beginning to a great new enterprise: to assure the survival of most Americans and certainly of the political, social and economic fabric of

American society under any and all circumstances of nuclear attack?

In public statements, policy-makers have both endorsed and rejected the idea that nuclear war could be won. The strategic modernization program could be characterized as movement towards the acquisition of the capability to wage and win nuclear war. No matter what is said for public political consumption, is the United States developing forces, and seeking to be able to protect American civilian, with a view to attaining the policy option of being able to press a military conflict to the point of victory? Or, is the Administration simply intending to arm the United States so that it can fight creditably?

The questions specified above point to a certain vagueness at the intellectual heart of the U.S. policy process, to an absence of authoritative doctrine. This appears as a problem with reference to virtually every major strategic program. The fragility, or even plain absence, of doctrine makes it very difficult for adequate, plausible policy stories to be told in support of proposed programs. For example, an important reason why the MX basing controversy has achieved its saga-like aspects, is because officials have never had a robust and authoritative story to tell in generic defense of land-based ICBMs and in specific defense of land-based ICBMs with the technical/strategic qualities of the MX. The story could easily have been written but never was.

No administration, from the early 1960s to the present day, has had either the ability or the political resolution to defend civil defense with a coherent and plausible strategic rationale. If an Administration cannot decide whether or not it wants area BMD capability, it is necessarily difficult for the candidate technical program to be considered in a strategically rational manner. The Reagan Administration believes that an American capability to wage protracted war is important for deterrence (pre and intra war), and may well serve vitally to limit damage. But, what are the details of the policy guidance that should instruct the "working levels" of government? Everybody favors the survivability of forces and C^3I, but how much endurance is necessary or strongly desirable? In addition, what kinds of military capability should be able to endure?

THE SPACE-BASED LASER CONNECTION

Policy trends are clear enough, but much of the guidance for implementation remains uncertain, as do the objectives toward which trends should aspire. Detailed analysis of the potential possible relevance of space-based laser weapons for the support or execution of strategic missions is beyond the boundary set for this chapter. However, it is perhaps useful for some candidate tasks to be identified as an important preliminary to a full investigation. What follows is a terse summary of space-based laser missions that are compatible, to varying degrees, with the strategic policy trends discussed here. This itemization is intended to be suggestive, not exhaustive. No attempt is made to identify down-side considerations that might weigh against laser deployment, and no measure of advocacy of deployment should be imputed to this discussion. This listing simply specifies ways in which space-based lasers could contribute to fulfill the missions established by strategic policy guidance. Deliberately, the list includes both generic missions, and particularly important detailed sub-sets of those missions.

Space-based laser weapons might:

1. defend U.S., and threaten Soviet, space-based C^3I assets;

2. provide the highest layer of multi-layered BMD system, for hardpoint, air and SSBN base, C^3I node, NCA and selected urban/industrial defense;

3. provide defense against high-flying manned bombers/strategic weapon carriers;

4. buy time for LUA or for a second-strike by denying the U.S.S.R. the ability to conduct "pin-down" attacks;

5. deny or contest the U.S.S.R. ability to conduct space-based or high flying post-strike reconnaissance, thereby grossly imparing Soviet ability to wage protracted war;

6. defend U.S. post-strike reconnaissance vehicles, satellites or air-breathing, thereby promoting U.S. ability to conduct protracted war;

7. provide BMD against any "third party" attacks.

This short list, embracing virtually every mission or subsidiary mission important in current official U.S. deterrence thinking, rests upon no precise technical assumptions beyond the promise that a first generation weaponized laser system in space would be able to achieve modest success. The most appropriate policy perspective upon space-based laser weapons is to view them as useful additions to the whole arsenal of U.S. war-waging capabilities. Proponents of lasers who tout them as wonder weapons capable of winning wars may be proved to be correct, but the long history of warfare suggests otherwise and, moreover, such extravagant claims place a burden of performance expectation that would be unreasonable to ask of a new weapon family that must face countermeasures currently of very uncertain effectiveness. In short, the possibility of being very substantially in error in one's prediction concerning the likely strategic utility of space-based lasers is so large that extremes of optimism or pessimism are both factually unsupportable at present.

What is most important now is that the potential relevance of space-based laser weapons be recognized unambiguously, and that strategic policy constantly be reexamined as the technological story evolves.

SUMMARY

The pace and direction of a U.S. space-based program must be seen against the background of the developing strategic policies of the Reagan Administration and its immediate predecessors. Since adoption of National Security Decision Memorandum (NSDM-242) in 1974, the Department of Defense has taken important steps to direct the strategic nuclear targeting policy of the United States away from an emphasis on large-scale countervalue targeting of the Soviet Union toward greater provision in the SIOP for more limited kinds of strike options executed against military and military-related targets, the intention being to provide the NCA with a greater degree of flexibility in carrying out central war plans. Presidential Directive (PD) No. 59 adopted during the Carter Administration sustained the thinking and basic policies contained in NSDM-242, but gave greater emphasis to the possibility of a protracted nuclear war, requirements for strategic force endurance, the deterrence value of holding at risk the Soviet decision-making elite, and the requirements for a secure strategic reserve force. The Reagan Administration essentially is focusing at present on implementing these earlier policies and taking a closer look at certain issues which had not been satisfactorily resolved earlier, e.g. the launch-under-attack

(LUA) option, C^3I requirements, and the force-structure implications of a protracted nuclear war. A master plan has been developed which is intended to relate strategic nuclear doctrine and targeting policy more systematically to the weapons acquisition process.

The Reagan Administration also has inherited from earlier administrations the problem of ICBM survivability and has been compelled to accelerate its decision-making with respect to the ultimate basing mode of the MX missile.

ICBM survivability and the policy revisions reflected in current nuclear employment policy are the two decision areas at the strategic level where space-based weaponized lasers have near-term, if not immediate, relevance. Such lasers are under study in the context of a multi-layered BMD system where they would be deployed to intercept Soviet ICBMs in the boost-phase of their trajectories prior to release of the RV bus. Space-based lasers are also of interest in connection with ensuring the survivability of U.S. space-deployed systems of all kinds.

Space-based weaponized lasers could make an important contribution to a U.S. damage-limiting strategic policy by defending assets other than those associated with strategic force survivability and the prosecution of a nuclear war, e.g. the defense of cities. But there is, unfortunately, no clear indication that the Reagan Administration is prepared to commit itself unequivocally to that school of thought. That prospects for such a commitment depend heavily upon the technological progress demonstrated in BMD research and development.

It should also be noted that the Reagan Administration has been criticized by members of Congress for not developing a coherent strategy, a strategy that would give clear purpose to the acquisition of weapons being sought by DoD planners. At the present time there is an absence of authoritative strategic vision informing the on-going militarization of space: until such a vision materializes, one cannot be sanguine about the coherence of future U.S. space defense policies (R&D, deployment policy, employment doctrine and arms control).

It is clear that space-based laser weapons could make a dramatic contribution to U.S. national security, but one should avoid casting these weapons in the role of wunderwaffe. They should be seen as a critical component in the spectrum of weapons deployed for deterrence.

Acronyms and Abbreviations

ABM	anti-ballistic missiles
ACDA	Arms Control and Disarmament Agency
ALCM	air-launched cruise missiles
ASAT	anti-satellite weapon
ASW	anti-submarine warfare
BMD	ballistic missile defense
C^3	command control and communications
CD	U.N. Committee on Disarmament
CONUS	continental United States
CPSU	Communist Party of the Soviet Union
CSB	closely spaced basing
DARPA	Defense Advanced Research Project Agency
DDR&E	Director Defense Research and Engineering
DEW	direct energy weapon
DIA	Defense Intelligence Agency
DoD	Department of Defense
DSAT(s)	defense for satellite(s)
DSMP	defense meteorological program
ECM	electronic countermeasures
ENMOD	Environmental Modification Techniques
ERCS	emergency rocket communication system
ER	enhanced radiation
FY	fiscal year
FYP	five year plan
GLCM(s)	ground-launched cruise missile(s)
H°	hydrogen atoms
HOE	homing overlay experiment
ICBM(s)	intercontinental ballistic missile(s)
INF	intermediate range nuclear forces

IOC initial operation capability

LoAD(s) low altitude defense system
LODE large optics demonstration experiment
LOW launch-on-warning
LWIR long-wave infrared

MAD mutual assured destruction
MARV(s) maneuverable reentry vehicle(s)
MEV million elechon volts
MIRV(s) multiple independently targetable reentry vehicle(s)
MPS multiple protective shelter
MRBM medium range ballistic missiles
MT megaton
MX developmental American ICBM

NATO North Atlantic Treaty Organization
NCA National Command Authority
NNK non-nuclear kill
NORAD North American Air Defense

PD Presidential Directive
PRO anti-missile defense (Soviet)
PVO anti-air defense (Soviet)

R&D research and development
RDT&E research, development testing and engineering
RV reentry vehicle
RW radiological weapon

SAC Strategic Air Command
SALT strategic arms limitation talks
SAM(s) surface-to-air missile(s)
SBL(s) space based laser(s)
SCC Standing Consultative Commission
SIOP Single Integrated Operation Plan
SLBM submarine launched ballistic missile
SLCM submarine launched cruise missile
SRAM short-range attack missile
SRF strategic rocket force (Soviet)
SSBN nuclear ballistic missile carrying submarine
START Strategic Arms Reduction Talks

Index